Test Economics and Design for Testability

for Electronic Circuits and Systems

ELLIS HORWOOD SERIES IN ELECTRICAL AND ELECTRONIC ENGINEERING

J. Beynon	**PRINCIPLES OF ELECTRONICS: A User-Friendly Approach**
R.L. Brewster	**TELECOMMUNICATIONS TECHNOLOGY**
R.L. Brewster	**COMMUNICATION SYSTEMS AND COMPUTER NETWORKS**
S.J. Cahill	**DIGITAL AND MICROPROCESSOR ENGINEERING: Second Edition**
C. Dislis, J. Dick, I.D. Dear, A.P. Ambler	**TEST ECONOMICS AND DESIGN FOR TESTABILITY**
A. Klos & P. Aitchison	**NETWORK SOLUTION: theory and practice**
J, Manyika & H. Durrant-Whyte	**DATA FUSION AND SENSOR MANAGEMENT**
P.G. McLaren	**ELEMENTARY ELECTRIC POWER AND MACHINES**
P. Naish & P. Bishop	**DESIGNING ASICS**
J.R. Oswald	**DIACRITICAL ANALYSIS OF SYSTEMS: A Treatise on Information Theory**
J. Richardson & G. Reader	**ELECTRICAL CIRCUIT ANALYSIS**
J.N. Slater	**CABLE TELEVISION TECHNOLOGY**
J.N. Slater & L.A. Trinogga	**SATELLITE BROADCASTING SYSTEMS: Planning and Design**
R. Smith, E.H. Mamdani, S. Callaghan	**THE MANAGEMENT OF TELECOMMUNICATIONS NETWORKS**
E. Thornton	**ELECTRICAL INTERFERENCE AND PROTECTION**
L.A. Trinogga, K.Z. Guo & I.C. Hunter	**PRACTICAL MICROSTRIP CIRCUIT DESIGN**
Wen Xun Zhang	**ENGINEERING ELECTROMAGNETISM: Functional Methods**

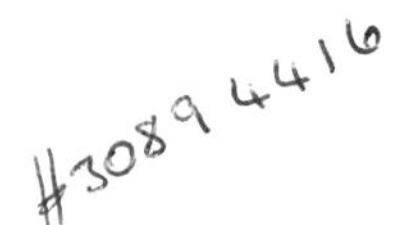

Test Economics and Design for Testability
for Electronic Circuits and Systems

C. Dislis
J.H. Dick
I.D. Dear
A.P. Ambler

ELLIS HORWOOD

New York London Toronto Sydney Tokyo Singapore

First published 1995 by
Ellis Horwood Limited
Campus 400, Maylands Avenue
Hemel Hempstead
Hertfordshire, HP2 7EZ
A division of
Simon & Schuster International Group

Printed and bound in Great Britain by
Bookcraft, Midsomer Norton

Library of Congress Cataloging-in-Publication Data

Available from the publisher

British Library Cataloguing-in-Publication Data

A catalogue record for this book is available from the British Library

ISBN 0-13-108994-3

1 2 3 4 5 99 98 97 96 95

Contents

Preface

According to a famous study by Robert Solow, over 80% of the growth in output per labour hour in the United States during 1909-1949 was due to technical progress.

Other significant studies have also shown that the growth in output cannot be explained entirely on the basis of the growth of capital and labour. Research, education and training contribute to advances in knowledge and efficiency which are the hidden sources of growth.

Technical progress influences the society in more than one way. First, technology helps us improve products and manufacturing processes. Electronic systems used in communication, computing and entertainment are good examples of such improvements that we have seen in recent times. The second aspect of technology relates to the introduction of entirely new products. Again, we notice that electronics has brought a more than noticeable change in our life-style in this century.

Coming to the main topic of this book, we find that testing plays a major role in the life of an electronic product, or for that matter, any product. The product is first tested when manufactured. It is repeatedly tested throughout its useful life for maintenance, diagnosis and repair. How does then the so-called 'design for testability' influence the cost? Take product A which is specially designed for testability. Its manufacturing test is very thorough and so the product is trouble-free and reliable. It is easier to diagnose and repair if it were to fail during use. Of course, we may have a cheaper product B which is designed without the stated consideration for testability. Product B fails more often, takes more time to repair, and causes loss to whatever business its user is in. Based on my description of the situation, you are inclined to choose A over B. But, does it make economic sense? That is, have you considered the price (the fixed cost), the operational cost (variable cost) and the period of intended use to compare the total cost? At times, your decision may differ depending upon whether the product is to be used for a short time or a long time.

Developing economic sense is essential for engineers who strive to design and produce the best possible products. There is always a cost associated with any improvement - testability is no exception. The engineer should weigh the cost against benefits in the intended application. Most engineers, like myself, are much better equipped with scientific details than the knowledge of economic principles. It is for this reason that I welcome this book. Tony Ambler and his co-workers, Chryssa Dislis, Ian Dear and Jochen Dick are pioneers in the area of test economics, and have widely published their original contributions. Fortunately for us, now they have put this valuable information together in one volume.

This book deals with economics based test strategy planning. The authors discuss specific design for testability methods like scan and built-in self-test. Integrated circuits as well as printed circuit boards are considered. Both design engineers and business managers will find the information on economic modelling and decision-making tools to be very useful. I hope this book will get to everyone's shelf so that we start reaping the benefits, economically.

Vishwani D. Agrawal
AT&T Bell Laboratories
Murray Hill, New Jersey

1

Introduction

1.1 DESIGN FOR TESTABILITY AND THE COST OF TESTING

When design for testability (DFT) became a term that everyone recognised, it did not necessarily follow that everyone would even consider the implementation of any such methodology in their own circuit designs. It was something that was discussed at conferences, usually by academics or industrial research centres, but rarely was implemented in *real* circuits. One obvious exception to this was in the case of IBM and their use of the scan path methodology for the systems that they designed and manufactured.

One of the reasons often quoted for not implementing DFT in an integrated circuit was the increased circuitry that was required - overheads of 10% were quoted, sometimes more, sometimes less. But this would prove too costly to implement, wouldn't it? Today, with a more realistic attitude to silicon costs ([Needham91] quotes a figure of 0.001 cents per gate for a 'moderate sized' device), overheads in excess of 20% can be quoted without anyone getting reluctant. The reality of increasing quality requirements for ever larger circuits has led to the necessity to use DFT methods - no other realistic solution currently exists.

So why are so many design teams apparently still reticent to use DFT? There will be situations where DFT cannot be used for technical and commercial reasons. A company that intends to push current technology for both functionality and performance in order to gain a competitive advantage over a competitor will not always be able to find the space for DFT. At the same time, it must be said that this might be seen to be a short-sighted strategy as the device/circuit produced will not be testable to the ultimate extent and a possible recall due to an obscure defect may be necessary - it is also said that a useful feature of DFT is in its design debug capability which leads to an improvement in time to market.

Many will now argue that test 'costs' should not be a consideration, that test should be as much a part of the specification as basic functionality. Ben Bennetts [Bennetts92] introduced the term quality improvement factor (QUIF) as a value that should be used to determine the benefit or otherwise of any DFT method, and he went on to suggest that the word 'overhead' should be banned. Others put the point another way by asking why test as a cost is discussed at all - shouldn't test be considered as much a part of design specification as functionality and performance? But still the question appears to be one of 'what value is added to a circuit/product through the use of DFT'?

Many economic analyses have demonstrated that the introduction of DFT with all its attendant 'overheads' will still lead to a higher quality product at *reduced* cost, assuming that a high quality product was wanted in the first place - it can be argued that if you don't want a quality product, then you don't need to test, so no test costs, therefore cheaper! But such a naïve attitude to quality neglects other issues, e.g. perceived value by customers and future sales, warranty claims, possible litigation by customers and users through damages caused by product failure. In many cases now, customers are requiring from vendors penalty clauses in purchase contracts for down-time that exceeds a set limit, and even for 'consequential loss' - payments for loss of revenue caused by the non-performance of supplied equipment. If such failures are caused by inadequate testing, then the economic issue takes on a whole new meaning!

However, the problems associated with test economics are not restricted to disputes between pro-DFT acolytes and anti-DFT adherents. For example, there is an argument raging at the time of writing between those who propose a partial scan design methodology and those who propose a full scan design style. The arguments surrounding the discussion are many and varied but usually will include, among others, performance impact, design time and fault coverage. Ultimately, provided that time scales are met, that functionality and performance criteria are met, that space limitations are not exceeded, that fault coverage requirements are at least equalled, the major requirement must be on the basis of overall cost to the company.

It can be argued that not enough information is available that relates to the economics of electronic circuit design, either in terms of specifically design-related issues or test issues. There are exceptions to this. [Davis94] is a book totally given to test costs (specifically oriented towards automatic test equipment), and there is material related to design costs of integrated circuits in [Fey88a] and [Fey88b]. More recently there has been a series of International Workshops devoted to the topic of Economics in Design and Test. Figure 1.1 shows a trend of costs against test-related area overhead which makes certain assumptions, but which also creates certain unanswered questions, e.g. what level of area overhead leads to the minimum cost? Other material can be found in other papers and publications but it tends to be sparse. However, what there is can be illuminating and interesting.

Nevertheless, what information there is must be studied carefully for it cannot relate, other than in a very broad sense, to any one set of individual circumstances. The range of factors that can affect test costs can vary widely from one installation to another, from one design to another. However, it is often the case that an economic analysis of individual test costs will point the way to more efficient use of test tools and eventually relate to improved product quality when it is realised that good test can pay back.

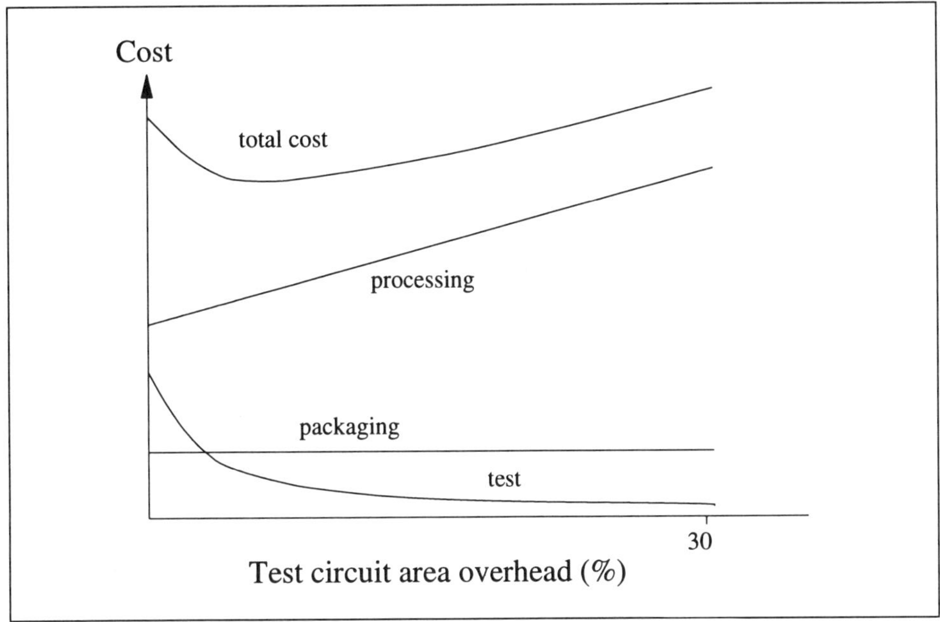

Figure 1.1 Cost versus test-related area overhead

The material in this book discusses various issues in the area of test costs leading to the development of test economics models that were originally developed as pure spreadsheet models, and are now being used to drive test strategy planning tools.

1.2 ISSUES IN TEST ECONOMICS

As might be expected, the cost of test, if nothing is done to ease the problem, increases faster than linearly for a linear increase in circuit complexity [Turino90]. The same source also states that both design verification and test program generation time can be cut by 5-15% if the circuit is made testable from the outset. At the same time, while total product cost has decreased over the past few years, test cost has risen as a percentage of this total product cost to in excess of 55% in some cases - dependent upon the product size, technology and complexity.

Unfortunately, there are also many myths surrounding test costs. It is easy to see that test pattern generation has not been generally accepted as a possible answer when observed from a fault coverage perspective, and from a CPU time point of view. The fact that significantly large CPU times can be quoted for fault simulation is often used as a reason why effective test is not possible. But equally well, although there are a variety of solutions to these difficulties, their acceptance has not been as widely supported as might be expected. The reasons given for the lack of use of DFT methods, scan and built-in-self-test (BIST), are plentiful but can include the familiar problems of area overhead, pin count overhead, etc.

The end result of these misconceptions can be that many devices/systems are inadequately tested. Note the statements by many silicon vendors that up to 50% of 'working' application specific integrated circuits (ASICs) do not work in the target system because the designs have neither been specified nor verified correctly through lack of simulation [Huber91], and have not been tested adequately. The use of design verification tests for device test typically provides 40-70% fault coverage. This is despite the warnings now stated widely that in excess of 98% fault coverage is necessary to ensure an adequate level of quality, not forgetting the six sigma requirements many companies adopt, which aim not only for a very high quality, but also for a very small deviation, which translates to a target yield of 99.99966%.

To the uninitiated, publicity that shows the cost of test rising dramatically can only serve to reinforce their opinions. Concerns as to the real cost of test mean that short-cuts are inevitable, leading to a naïve reliance upon the results of perhaps questionable test generation and fault simulation methods.

The 'cost' of test, being a negative quantity, is far easier to focus on than are the benefits, especially when the 'over-the-wall' mentality prevails, i.e. when the design team is asked to use its resources to improve the lot of the test department with its own budget. Why should the design team spend more of its money in order to produce cost savings to the benefit of the test department?

However, in order to make realistic choices, the problem of testability provision has to be examined in more detail, especially from the perspective of the whole company, not just of individual departments.

As an example of a consideration of test costs, take the use of scan path. The use of scan path as a test methodology is well known and will not be expanded upon here since reference can be made to many publications, e.g. [Abramovici90].

The benefits of using scan are easy to define and include

- increased observability and controllability
- reduction of the test problem from one of sequential circuit test to that of a combinational circuit plus a shift register
- automatic test pattern generation (ATPG) capability being available commercially for combinational logic, giving the possibility of near 100% stuck-at fault coverage
- the fact that it can be used at all stages in a system design, from initial design debug, device production test, to system test and field service.

However, there are the often cited penalties of using scan, including

- additional design effort required
- additional circuitry required (4-20% overhead)
- additional device pins required, which can result in the necessity for the next package size - space, price
- possible performance degradation
- possible increase in test application time

- reliability and yield impact

An industry survey [Racal89] reports that very few ASIC designs incorporate scan; the reasons often quoted are those of area overhead, degradation in performance, reliability and yield, and the necessity for design constraints [Wilkins86].

Some of the negatives above relate to design philosophy and performance issues which will also not be discussed here. Suffice to say that there is a wide range of software tools and support available for scan design including design rule checkers, automatic scan conversion tools and automatic test pattern generation.

The other issues are more directly cost-related in nature and require some detailed analysis for a true comparison to be made. The factors that are involved here will include, but should not be limited to:

- the balance between increased gate count and the impact this might have of increased design time, increased processing costs (less devices per wafer that can be made), decreased yield and reliability
- the balance between easier test pattern generation/higher fault coverage, and potential increase in test time (serial application of tests, which might be aggravated or eased by the absence or presence of a scan tester).

All these factors can be compared on straight economic grounds and would provide for a quantitative evaluation of scan as a test method - the qualitative issues are for the individual design teams to evaluate.

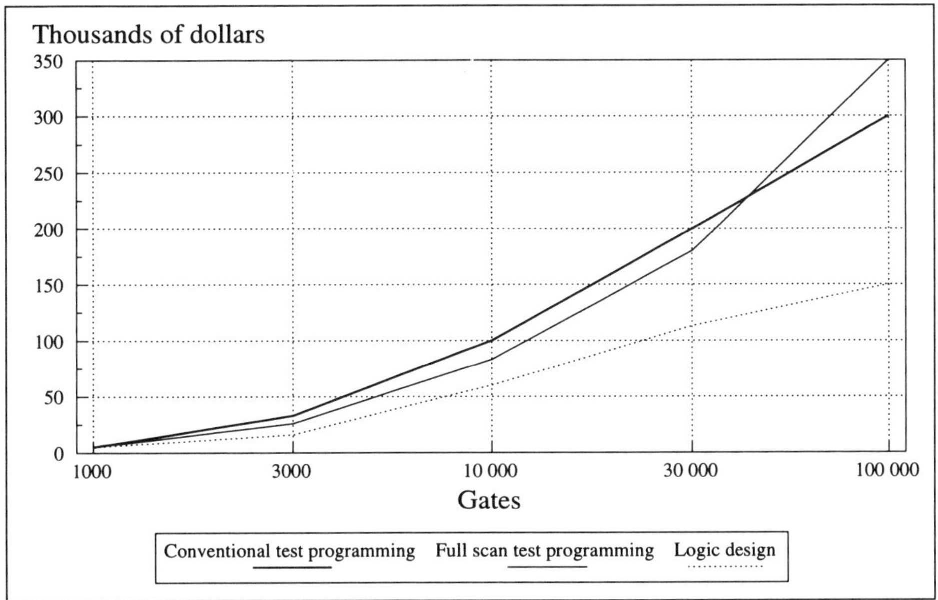

Figure 1.2 Scan costs

Nevertheless, it has been shown that the cost of test, with or without scan, is very much in favour of using scan [Racal89], (see figure 1.2), and this does not take into account the benefits to the board, system and field test stages were the entire design to include the scan design philosophy. [Dervisoglu88] demonstrated the benefits of system-wide scan test.

The effectiveness of improved testability, and hence scan, at the device level to subsequent test stages is simply put by reference to the 'rule of tens' [Davis94], illustrated in figure 1.3.

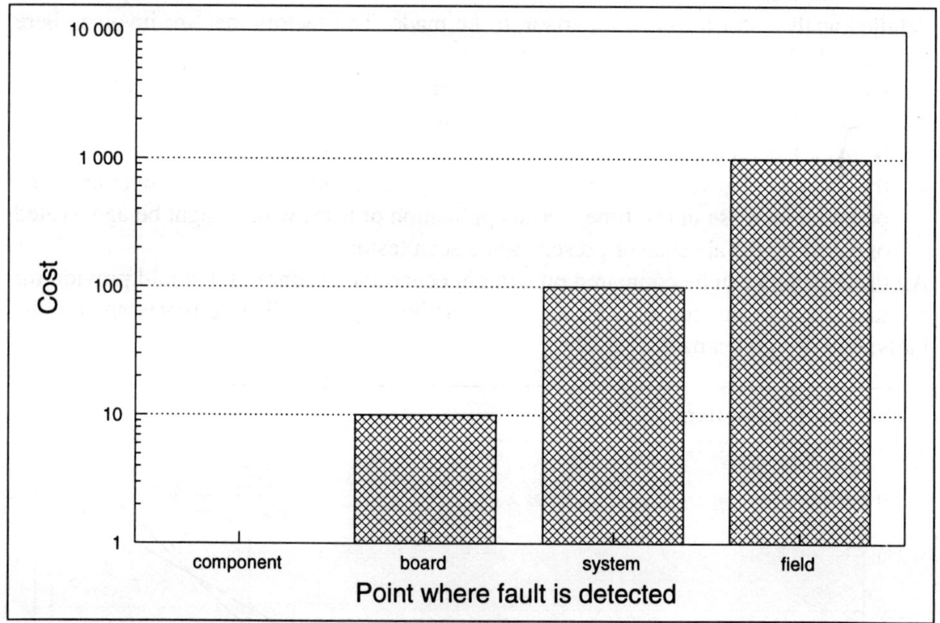

Figure 1.3 Rule of tens

This demonstrates the effects of leaving test to the later stages of system integration and graphically shows that poor quality test at the device level will lead to very large increases in costs when the effects of this are felt in field failures.

In short, if the cost of detecting a fault at the device level were to be, say, $1, then to detect that fault at the board level would be $10, at the system level it would be $100, in the field $1000, etc. Although used as a basic rule of thumb, analyses of the real costs based upon actual examples have shown this to be a reasonable rule to follow (the figures for dealing with testing out a design flaw could follow a similar pattern, and be a lot higher).

Although a simple rule to follow, and therefore useful, more recent studies have shown the rule of tens to be inadequate in representing the effects of test costs in subsequent levels of integration taking account of the quality of devices that are produced in a modern production environment. [Davis93] describes the situation that now prevails, (figure 1.4). In the case of figure 1.3 the low cost of component test is justified by the relatively high numbers of defective devices that were present in any batch, so cost of test per defective device could be amortised quite effectively. In the case of figure 1.4, representing a modern scenario where defective devices are much lower in number, there is much less room for the amortisation. Just to cloud the picture further, other analyses show yet another outcome [Dislis93]. This helps to emphasise the problem with continuing to assume test-related costs even by the so-called 'test literate'.

But this also shows the problem with explicit use of 'rules of thumb'. It is too easy to place importance on them without thinking of the consequences for the problem to which they were applied and in those particular circumstances. Just how accurate were they in the first place?

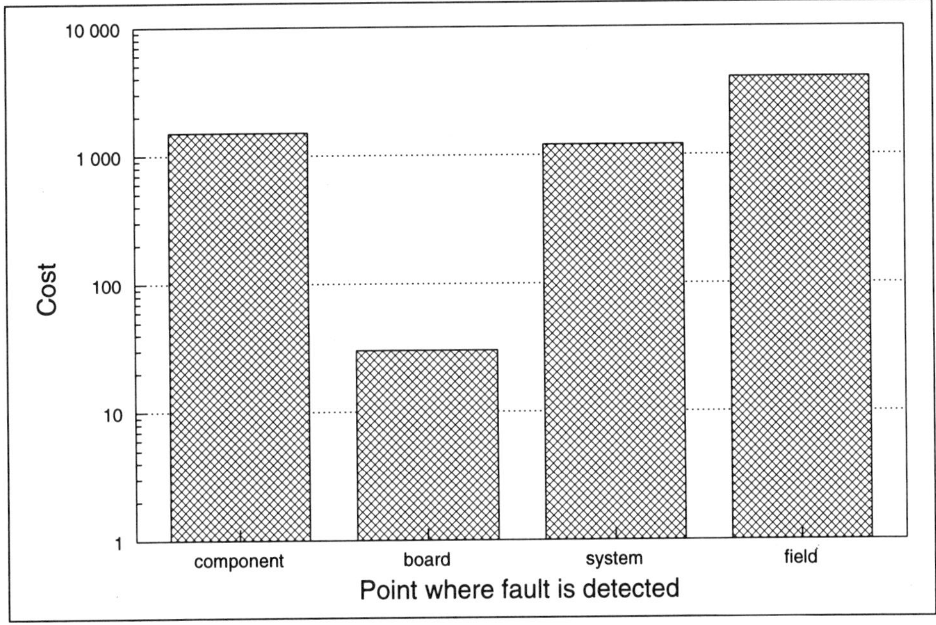

Figure 1.4 Rule of tens (updated)

Time to market is another important consideration in the development of a new product. The time to market for a new product can be considerably reduced through the use of DFT techniques. Adopting design for testability methods can significantly reduce development times and therefore cut the time to market. Design verification and test program generation can be cut by, typically, between 5% and 15% [Turino90]. Figure 1.5 provides an illustration of the time savings that can be made at different stages of the development of a product.

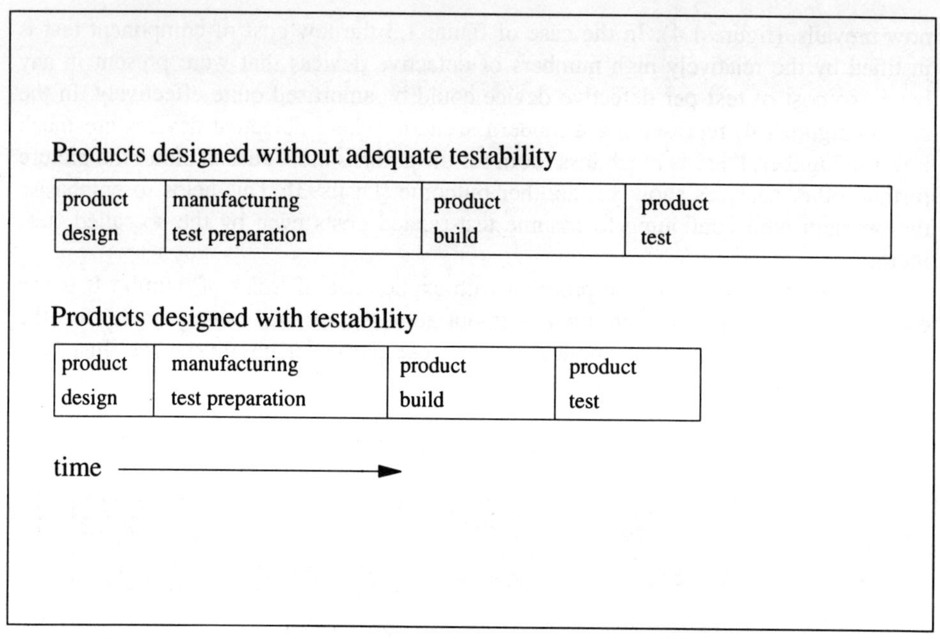

Figure 1.5 Development time comparison [Turino90]

Fast development time has real financial benefits for the manufacturer, although the exact economic advantage depends on the state of the market and the current competition [Smith91]. In general, early introduction of a product means that the product sales window is increased. If a product is introduced early, it seldom becomes obsolete any sooner, and the early start ahead of the competition also provides the manufacturer with more pricing freedom. The early introduction can result in a larger customer base and an increased market share, particularly in the case where switching to another manufacturer's product can be an expensive exercise for the customer. Figure 1.6 shows the financial benefits of operating in a longer sales window [Smith91].

1.3 NEED FOR ECONOMIC ANALYSIS AND COMPARISON

In addition to there being a need for a straightforward economic analysis of the costs of test, consider the dilemma of a designer who needs to choose a design for test strategy for his/her ASIC. Even for those that are *au fait* with the test techniques that are available, the choice can be a not inconsiderable one. For example, in [Zhu88] 20 built-in self-test methods for programmable logic array (PLA) test are discussed. Given that designs can be made up of structures other than, but can include, PLAs, then the list can become daunting.

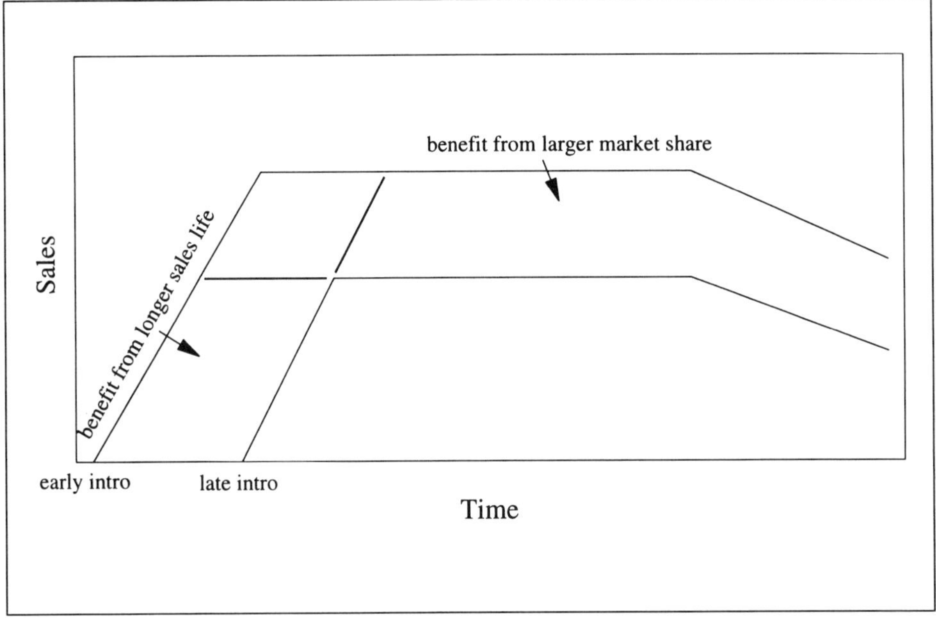

Figure 1.6 The benefits of early product introduction

Factors that must be considered when deciding upon the test method to be adopted in any design will include, but will not necessarily be limited to,

- area overhead
- pin count overhead
- test length
- fault coverage
- automatic test equipment (ATE) requirements

as well as design-dependent factors such as

- design time
- circuit size
- performance impact
- test pattern generation costs.

A fuller discussion of the parameters necessary is given later in this book, and in [Varma84], [Dislis89, Dislis 91, Dislis 93].

All of these factors will vary with circuit style. Some procedure is necessary to compare and contrast these disparate values in a way that can make a meaningful judgement. When confronted by what could be conflicting issues of longer test time versus larger area overhead versus increased pin count requirements, among many others, how can a *satisfactory* way of reconciling these be found?

Of course, there will be many cases where the variables listed are not to be considered as variable, e.g. there may be design constraints whereby area overhead or pin count is fixed due to package size limitations that cannot be changed (as in an aerospace example). However, the issue still remains as to how to compare and contrast effectively the effects of all the variables associated with a test strategy choice.

[Varma84] showed the effects of an analysis of many differing test-related parameters by using monetary cost as the basis for comparison. Graphs of specific cases showed how the relative benefits of BIST versus scan versus no DFT varied with production quantity. In the cases studied, the changes were quite marked, with BIST being the preferred method for lower production volumes, and no DFT for the higher production volumes.

The outcome of this work was to show that different test-related quantities can be effectively compared to each other using cost as the basis for comparison, allowing for true quantitative comparison. It would appear that any other form of comparison can only be subjective.

1.4 COST EFFECTS OF TEST RELATED PARAMETERS

The effects of adding testability to a design are widespread. To assess the impact of a design for test strategy it is important to consider both the direct impact on test costs (such as test pattern generation) and factors such as the reduction in ATE complexity and increase in design time and cost. This section will briefly consider some of the main tangible cost areas that are affected by test decisions, namely design, manufacture, test generation, test application. When costs are discussed it is important to remember that they can vary greatly between companies and products. The following is meant as a guide to the possible benefits or penalties of design for test.

1.4.1 Area overhead and production cost

The area overhead penalty for design for test is often quoted as one of the main arguments against design for test. The argument of increased chip area because of DFT resulting in decreased yield and hence increased production cost is well understood. However, there is not a general figure that can be placed on the maximium allowable area overhead associated with DFT. The designer is often left to the best assessment available, or a limit is imposed on the design from another source. The origins of this limit are often hard to ascertain and the question, ‘has the cost impact of the area overhead been properly considered?’, is often left unanswered.

Most companies in the fabrication industry keep a close eye on the chip yield of their process lines. Depending on the type of device produced, the technology, etc., these companies will have models to calculate the yield of devices, given a chip area. There is an abundance of research activity devoted to this field. Even if the designer had access to these models he/she would need to calculate accurately the area overhead of the whole chip before the effect on the yield and subsequently the effect on the production cost could be assessed.

In some cases the complexity of the analysis required can be reduced considerably. It is not uncommon for main companies to contract out the manufacture of their semi-custom chip design to specialist manufacturers. It is common practice that for a given production volume and package type the production cost per unit is based upon a simple gate count model.

It should be pointed out that any error on the part of the manufacturer in setting costing levels can be very dangerous. However, the manufacturer is often very experienced and has a good understanding of the economics. The actual costs quoted of course will vary from manufacturer to manufacturer. Obviously if a company places many designs and or large production volumes through the manufacturer then lower costs per unit can be expected.

1.4.2 DFT and design time

The design time increase for the addition of DFT is influenced by several factors. Fey [Fey87, Fey88a] produced an empirical model from studies carried out at the Xerox Corporation. The model calculated the efficiency of the design department based on a set of characteristics related to the design environment and the design itself. This work highlighted factors such as the familiarity of the designer to the design; originality of the design; type of technology; CAD tools available; as well as gate count affecting the design time.

The design time penalty of DFT methods is often quoted as an important factor because of its effect on time to market. However, if the design environment has been set up to encourage DFT and the designers are skilled in that area then the penalties are much reduced [Di Giacomo89, Turino90]. Taking this argument to the extreme, the use of a structured DFT approach can often increase the efficiency of the design centre and reduce the amount of rework required. Furthermore, it is the effect on design time and test generation time that is important. With modern design systems it can often be the test department that has the bottleneck.

1.4.3 Test pattern generation costs

Test pattern generation (TPG) cost is a one-off engineering cost. Thus the smaller the production volume the greater effect this cost can have. This is the main reason for the commonly quoted trend that increasing DFT becomes more attractive as production volumes fall [Varma84, Dear87]. The fall in production volumes can be linked to the increase in integration and subsequently increased testing costs [Di Giacomo89].

The calculation of accurate test generation costs is very difficult. However, it has always attracted a high level of interest. One of the first detailed studies of automatic test pattern generation (ATPG) was carried out by Goel [Goel80]. Goel's study proposed a relation between the number of test vectors (effort) and fault cover for ATPG and showed how test lengths can be associated with test pattern generation costs. Goel produced an empirical model but the application scope was limited to combinational blocks. However, this model was used by Varma [Varma84] to show that DFT methods still had an advantage when system and field test is considered. Varma reinforced his argument by adopting very optimistic figures for devices with no DFT (i.e. assuming combinational circuits for TPG predictions).

This type of model can be extended to take into account a scenario where ATPG is performed to obtain the maximum level of fault cover that is possible by specific ATPG systems for sequential designs. In such a model both the limitations of the ATPG package and the characteristics of the design need to be considered. As in the combinational case presented by Goel, the gate count, number of inputs/outputs and the average fan-in/out of the circuit can be a good measure of the circuit size and complexity. For sequential circuits the sequential depth and number of storage elements in the circuit can give a measure of the sequential nature of the circuit. The limitations of the ATPG system used will be affected by the algorithm used, amount of machine memory, skill of the operator, etc.

A similar method can be applied to manual test pattern generation. Manual TPG is normally required to increase the fault cover from the maximum achievable level obtained by the ATPG to that required for the design, which is often in the region of 99.9% fault cover of all *testable* faults.

Such a general modelling approach must be treated with care. It is important to use historical data consistently and understand the limitations of such a general modelling prediction to specific designs. Normally, with the use of DFT such as scan path or BIST a model can produce more accurate test generation cost predictions as the complexity of the circuit is often reduced. If pseudo-random or exhaustive tests are used then only fault simulation costs are required. In some cases fault simulation cost can be considerable. If 'canned' tests are used (i.e. pre-stored tests for commonly used circuit building blocks) then the cost of test generation and possibly fault simulation can be saved.

1.4.4 Test application cost

The test application cost is primarily affected by the number of test vectors per device, the number of devices and the type of ATE used. The type of DFT adopted can impact heavily on the number of test vectors as well as the type of vectors to be applied. In some cases the DFT can impose restrictions on the type of ATE used or vice versa. The argument put forward for the use of BIST is a valid one, i.e. the need for an inexpensive tester to act as control only for the self-test logic. However, the test time or throughput costs can still be the critical factor for high production volumes if exhaustive testing is performed by self-test logic.

There are many trade-offs that can be made between test lengths and DFT methods and there are individual examples to support. In general the type of system/device and the production environment influence which way the trade-offs swing. For example, scan path testing results in long sequences of test vectors to be applied of narrow width. The width depends on the number of different scan paths and the amount of parallel test application that is possible. Thus for scan path testing to be efficient the ATE requires a large memory behind some pins with very fast application speeds. Without this type of ATE available the application time/cost removes it as a possible DFT approach unless the production volumes are very small. The option of capital investment to purchase a new ATE is often a corporate decision and needs to be based on more than one design, [Huston83, Davis94].

The use of partial scan [Cheng90] can often remove the need for an expensive scan path tester and reduce the total number of test vectors to be applied, thus generating a cost effective test application solution. Also, the set-up time for probes etc. can be an important factor. If partial scan can reduce the problem of probing then there can be a significant saving in production test time.

1.4.5 An illustration

As an illustration of the test costs incurred and the trade-offs required, consider the following question: 'what is the maximum area overhead that can be afforded for a given test method?'. It was pointed out previously that the overall chip size affects the cost effectiveness of DFT methods for the circuit, due to reduced yield. However, in the test sector, there are larger savings to be made in a large circuit. The cost effectiveness of DFT then depends not only on the number of devices produced, but also on the size of the device as well as the area overhead of DFT methods contrasted with savings in test costs. To illustrate this point, several ASIC designs were evaluated. Gate counts and areas of the devices are summarised below.

1. 5000 gates 0.50 cm^2
2. 10 000 gates 0.80 cm^2
3. 50 000 gates 2.00 cm^2

The results of this analysis are shown in figures 1.7 to 1.9 [Dislis89].

A range of area overhead values was considered. The range for scan was between 5% and 15%, and for self-test between 20% and 30%. Costs were again normalised to the no-DFT case. The difference between the high and low overhead cases widens as the size of the chip is increased, due to reduced yield.

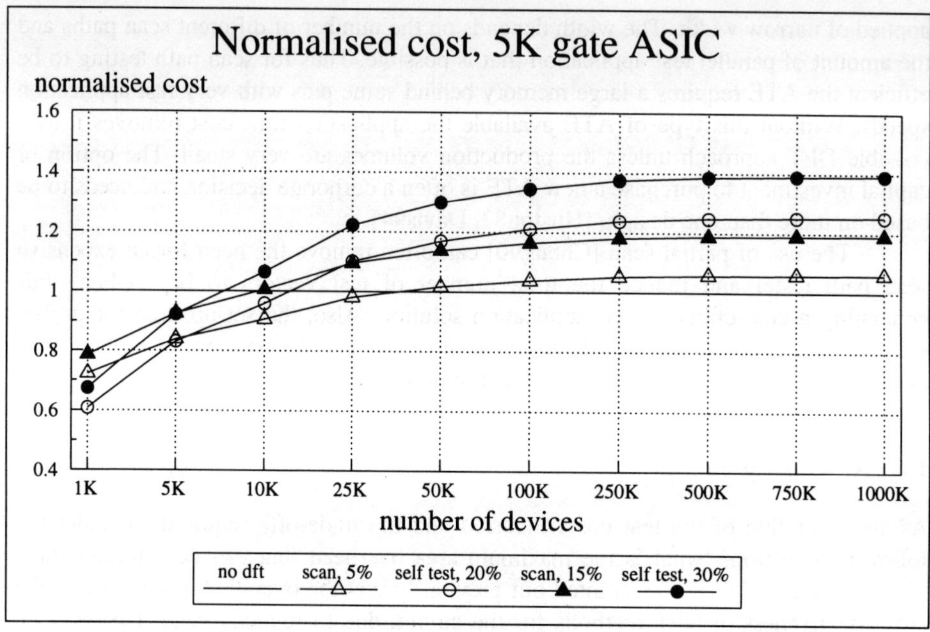

Figure 1.7 5000 gate ASIC

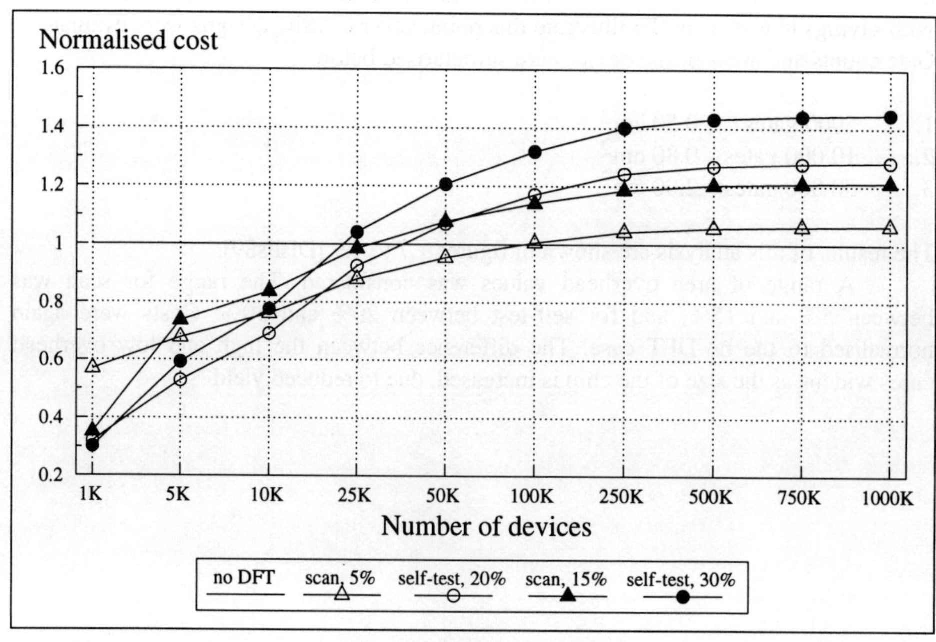

Figure 1.8 10 000 gate ASIC

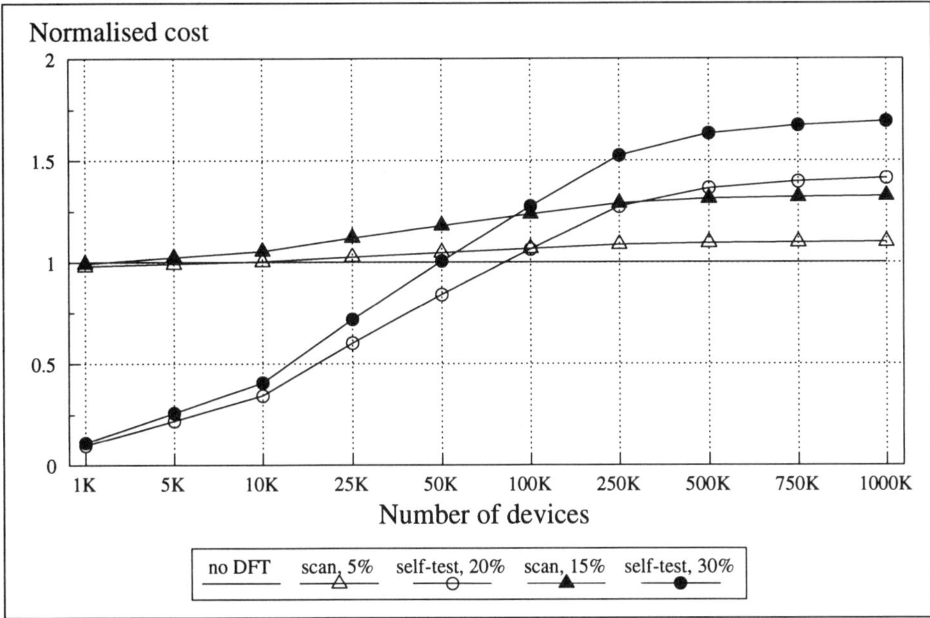

Figure 1.9 50 000 gate ASIC

In the case of self-test, the crossover point (the number of chips at which self-test ceases to be cost effective) increases as the gate count increases from 5000 to 10 000 gates, and is reduced again as the gate count is increased to 50 000. The reason is that as the complexity of the circuit increases, test generation costs also increase. However, due to the fact that the chip size is increased as well, there is a point where the increased manufacturing costs become higher than the savings in the test sector. In the case of scan path, this effect takes place earlier, as the savings in test costs are not as high as in the case of self-test.

1.5 RELATED ISSUES

A straight economic analysis of test-related factors can be a very useful and illuminating task. However, there are other factors that cannot, in the authors' opinion, be measured accurately enough, if at all, in any economic calculation, but that must be considered at the same time. One of these is performance, which is of primary concern and yet one that the authors have not been able to include objectively in a cost analysis.

The effects upon the quality image of the company *must* be a major force in any considerations being made relating to the testability of any product. If an economic analysis did show that, for example, 70% fault coverage was acceptable on cost grounds, consider the effect on the company image when that inadequately tested device causes the destruction of, say, a civil airliner. Clearly cost considerations can be far outweighed by other factors.

Marketing must also be considered in any cost-effect analysis; it has not been considered by the authors, but is addressed in [Miles91] and [Levitt89]. Such factors are important and can greatly affect the costs and likely productivity of the finished product, both in terms of defining the original specification, but also in terms of time-to-market issues. Any design change/enhancement, which might be as a result of testability improvements for example, can affect the time that it takes to get the product to market (see chapter 3). Alternatively, if the concept is considered the other way around, i.e. time to market is late because testability has not been designed in, figures suggest that shipping 6 months late can lead to a 33% reduction in potential profits in certain market conditions [Reinertsen83] - this is in addition to the negative benefits of not incorporating testability!

There is also the issue of whether or not a merchant chip manufacturer has any interest in providing improved testability that will be for the benefit of the user/customer. The economic issues in these cases are interesting. How much extra might/could the customer be charged for this new design 'feature'? An argument might be that pushing the bounds of technology at the time that a new device is being designed will preclude the incorporation of testability as it would reduce potential functional area, but then how the new circuit be adequately tested?

2

Test strategy planning systems review

2.1 INTRODUCTION

This chapter describes the requirements for and outlines the structure of a range of systems which are used for test strategy planning. Previous work in the area will be discussed, and in doing so, the problems which need to be addressed are described, together with the solutions adopted by different systems.

The terms 'test strategy planning' and 'test planning' are used inconsistently to describe a wide variety of planning actions, not all of which are relevant to the work described here. For example, a test strategy plan might be used to describe the order and timing of test vectors applied to a chip or board, the use of different automatic test equipment within the company, or the overall approach of the company to test, in terms of the responsibilities of designers and test engineers, use of internal and external test resources, target fault coverage and quality assurance.

In this case, however, the scope needs to be defined more specifically. The terms are used to describe the planning and application of structured design to test methods for an integrated circuit. It covers decisions on whether a single or hybrid strategy should be adopted, as well as deciding on the optimum level of test provision in a design. In every case, test strategies must meet technical specifications. Under this definition, a broad variety of systems can be described as test strategy planners, although they may differ in their approaches, broadness of their application and test method selection.

This chapter will first attempt to offer a classification of test strategy planning systems, moving from systems that consider only one type of strategy to systems which consider multiple and hybrid strategies. Each type of system will then be examined in more detail, with examples and discussions of published work. The use of knowledge bases will also be examined, with some existing systems considered in more detail. Finally, the test strategy planning approaches which consider economic analysis will be outlined, with emphasis on the differences with other systems with similar goals, and the reasons why it might be better suited to certain applications. The aim of this chapter is to give a description of test strategy planning work in a logical sequence, and show how this affected the progress of the economic driven test strategy planning work which is the subject of this book.

2.2 TEST STRATEGY PLANNING SYSTEM CLASSIFICATION

Test strategy planning systems can be classified into two broad categories. Single strategy systems use an optimised form of a single strategy such as scan path or built-in self-test (BIST). In partial scan, for example, only a subset of flip-flops are altered, in an attempt to reduce the area and performance overheads of full scan [Trischler83, Agrawal87, Cheng89]. In a partial scan strategy, the problem to be addressed is the selection method used to determine the minimum number of additional flip-flops required in the design to form the scan chain, while still guaranteeing a comprehensively testable circuit. A number of solutions to this problem will be discussed in section 2.3.1.

There are also systems that will make a given design self-testable. This can be achieved either by using techniques that overlay BIST on a design [Ambler86], or by using a selection of different BIST methods applied with some optimisation in mind [Jones86]. This approach leads to the development of hybrid strategy systems, where a selection of DFT techniques can be used according to need, with or without optimisation. This can lead to solutions more appropriate for the specific application, as factors like the architecture and partitioning of the design can be taken into account. However, the problem becomes more complex when the definitions and interactions of several types of strategy have to be considered simultaneously, and this is often an expert task. For this reason, a number of knowledge based systems have been developed for different test strategy planning applications. Some design synthesis tools also take advantage of the multiple test strategy philosophy to synthesise testable designs.

2.3 SINGLE STRATEGY SYSTEMS

Although this book deals mainly with test strategy planning using multiple strategies, single strategy systems will be examined here for several reasons. There seems to be a logical progression from single to multiple strategy systems, and then on to knowledge based systems. The examination of single strategy systems can provide useful insights into the way methodologies are applied, and the types of problem that can be encountered. The advantages and drawbacks of techniques used by different single test strategy systems provide useful pointers in the development of test strategy planners which deal with multiple strategies.

2.3.1 Partial scan path systems

Scan path [Williams83] is a well known and documented structured DFT technique which relies on internal storage devices that are reconfigurable into shift registers to increase the controllability and observability of circuit nodes. Variants of scan path include level sensitive scan design [Eichelberger77], a method adopted primarily by IBM, and random access scan [Ando79], which provides direct access to internal storage elements by using random access addressing for each latch. Partial scan path systems deal with replacing a full scan strategy with a partial scan, keeping the fault cover as high as possible. In partial scan, only a subset of flip-flops in the design are made scannable. Scan path can sometimes incur an area and performance overhead which has made the technique less acceptable among designers, despite the fact that when all factors are taken into account it can be economically viable. In addition, long scan chains result in long test vector sequences, although these can be avoided if more than one scan chain is used. Extra scan chains would require that extra pins or pin multiplexing be used. If only a subset of flip-flops are in the scan chain, then the drawbacks of full scan can be minimised, while still improving the testability of the circuit. Performance degradation can also be minimised if flip-flops which are part of critical paths are not selected for the scan chain. The primary aim is to achieve the highest possible fault cover with the minimum number of scannable flip-flops, although factors such as test length or test pattern generation time also need to be taken into account.

In doing this, several problems need to be addressed. The main question, of course, is which flip-flops should be made scannable. The second problem is achieving a guaranteed high level of fault cover, bearing in mind that partial scan introduces some problems in test pattern generation compared to full scan, such as how to test the non-scannable flip-flops. There are three main methods for the selection of the position of scan flip-flops. Testability analysis methods can be used to indicate problem nodes and select a minimal flip-flop set [Trischler83]. Alternatively, test pattern generator data can be used to identify suitable sites [Agrawal87], and these two methods are sometimes used in combination [Intelligen89]. Finally, it is possible to identify a flip-flop set by analysing the structure of the circuit looking for constructs which are likely to create problems for test pattern generation, without going through the process itself [Cheng89]. Examples of all three methods are given below.

A testability analysis method for selecting suitable flip-flop sites for partial scan was described in [Trischler83]. The testability analysis package SCOAP [Goldstein80] was used to calculate a testability measure for all flip-flops in the design. Based on that measure, flip-flops with the worst testability were selected for inclusion in the scan chain. No information was given on the testability cut-off point used. The savings in automatic test pattern generation were compared to the increased cost in silicon area. Based on that model, it was calculated that the maximal cost product reduction is achieved when 15-25% of flip-flops are included in the scan chain. The author pointed out that a proper economic analysis of the technique would need to take into account other factors such as yield effects (a reduction in yield would be expected as a result of area overheads), and the cost of additional pins, pointing to the need for a full economic model to answer the question 'How much DFT is necessary?'. The method assumes that the testability package used is adequate for the purpose. There is little or no optimisation of the exact placement of scan flip-flops - for example, it may be better to place scan flip-flops prior to testability 'blackspots'.

A system which used a combination of testability analysis and test pattern generation techniques was the scan element placement module of the Intelligen Test Pattern Generator [Intelligen89]. The system is based on a sequential test pattern generation algorithm, and scan elements, called T-cells, can be added to the circuit to improve its testability. T-cells are designed so that they can be added to any node, not just storage element, and are intended to impose minimal design restrictions.

The initial test pattern generation run is used to perform a dynamic testability analysis and create a list of nodes in order of testability. The testability of a node is simply a calculated from test pattern generation data. A file of the nodes in order of test pattern generation difficulty is created, from which the user can select a set for T-cell placement. The test pattern generation cycle is then run again, and the process repeated until the required fault cover is reached.

The system aims to provide an automated method of scan element placement, based on test pattern generation itself, therefore by-passing the problem of how well the testability analysis maps to the test pattern generation effort. However, the selection of T-cell sites is rarely optimal, as the selection process does not take into account the multi-site improvement from a single T-cell placement. It is often worthwhile for the user to refer to the circuit schematic for T-cell placement, using the recommended sites only as guidance. This is because the recommended sites are simply based on the test pattern generation run, without taking into account the structure of the circuit in any other way. Therefore there are cases where T-cells are more effectively placed before the recommended site, as a greater improvement is often possible. In that respect, the T-cell placement process cannot be termed automatic, although the guidance is very useful. The Intelligen producers quote the same fault cover as full scan path, but typically with only 3 to 8% of the scan cells. The disadvantage, as with all scan cell placement methods which use test pattern generation, is that the test pattern generation cycle has to be run before the design can be updated for testability. Although the fault cover cannot be guaranteed at each stage, it can eventually be the same as that of full scan. In the worst case this may require the inclusion of all storage elements in the scan chain through an iterative process.

Two methods for flip-flop selection for partial scan using a combinational test pattern generator were described in [Agrawal87]. These were developed on the basis that an automatic test pattern generator would be used. Because unscanned flip-flops are in the unknown state as far as a combinational test pattern generator is concerned (the inputs and outputs provided by them cannot be controlled and observed by the test generator), and therefore some faults cannot be detected by the test pattern generator and must be covered by functional vectors, only flip-flops which are rarely used by combinational tests should be excluded from the scan chain. The target faults for scan test generation are the faults which are not covered by functional (verification) tests.

The first such approach used a modified PODEM (path oriented decision making) algorithm [Goel81] to generate all tests for each fault in the target list. An option table is provided for the designer, of estimated fault cover versus flip-flop use. An example of an option table for four flip-flops (A, B, C, and D) and four undetected faults is given in table 2.1. This is generated by keeping a record of all flip-flops used by each vector (**t**) while generating all test vectors for each target fault (**f**).

Table 2.1 Option table example

Fault fi	No. of possible tests	Flip-flops used		
		t_{i1}	t_{i2}	t_{i3}
$f1$	2	none	A, B	-
$f2$	3	A, B	B, C	A, B, D
$f3$	1	A	-	-
$f4$	2	A, B, D	A, D	-

The objective is to select, for a given number of flip-flops, a set of test vectors which will cover the maximum number of faults. No more than one test per fault should be used. The full set of tests is processed to find the minimal set of flip-flops. This is done heuristically, based on the frequency of use of unscanned flip-flops by each test. Flip-flops are added and the fault cover is estimated until it reaches the target fault cover specified by the designer.

The above method, although resulting in good optimisation of the number of scanned flip-flops, is expensive in terms of computation and storage, especially for large circuits. This is due to the fact that all tests are generated for each target fault. An alternative method was also described, where only one test per fault needs to be generated. This is based on the distance heuristic of the PODEM algorithm. In order to accomplish path sensitisation or line justification, PODEM sets the primary inputs which are nearest to the site of the objective, i.e. have the smallest number of gates in the path. In the partial scan test generation, inputs and outputs of combinational logic which are derived from flip-flops are assigned a large distance, to ensure that the test generator avoids using input signals as much as possible. An option table of estimated fault coverage versus flip-flop usage is set up, and the flip-flops that are used by the test generator to achieve the target fault cover are made part of the scan chain. This approach is not as optimised as the former one, but is less expensive in terms of computation and storage, and no post-processing of tests is required. The authors report an advantage of around 40% fewer flip-flops than full scan for the same fault cover (typically over 95%). These methods use verification vectors as the starting point for the partial scan strategy, so their success depends largely on the quality of the verification vector set.

Test pattern generation methods do provide guidance for the placement of scan flip-flops, but may involve a large amount of computation before a set of flip-flops can be selected. An alternative method is to analyse the structure of the circuit in order to pinpoint the factors that make test generation difficult (i.e. to identify the circuitry prior to test pattern generation), and select a set of flip-flops to eliminate these. Such a method was presented in [Cheng89], and the aim was to eliminate reliance on test pattern generators as well as functional patterns, by directly reducing the complexity of the circuit for the test pattern generator.

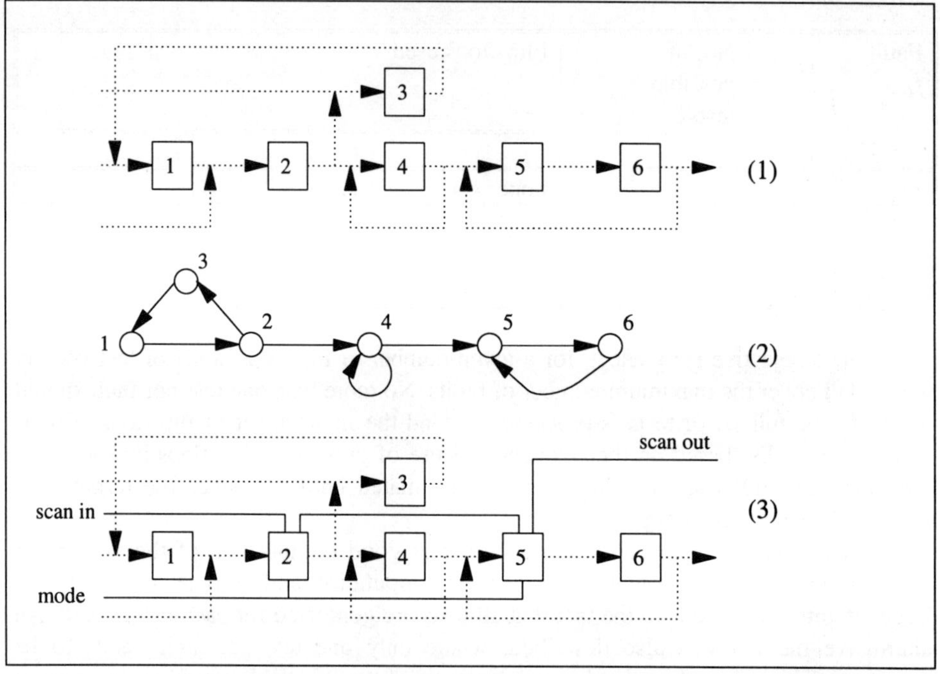

Figure 2.1 Selection of flip-flops to break cycles

The method is based on experimental data showing that the test generation complexity using a sequential logic test generator may grow exponentially with the length of cycles in the circuit, while it grows only linearly with the sequential depth. A graph theoretic method is then used to select a minimal set of flip-flops to break all cycles in the circuit. An example is shown in figure 2.1. Dotted lines represent signal flow through combinational elements. The circuit shown in (1) has six flip-flops, and the graph model of the circuit is shown in (2). It is apparent from the graph that by breaking the cycles at flip-flops 2 and 5, as shown in (3), all non-self-cycles in the circuit are broken. Flip-flop 4 was not chosen because it is a simple self-loop, and easily initialised.

The method takes into account the fact that loops may be embedded, and only attempts to break representative cycles, i.e. cycles which are not embedded in any other. Once a selection of flip-flops to break all representative cycles is made, a second algorithm selects extra flip-flops to reduce the sequential depth to a maximum, defined by the user. The authors also provided a separately controlled scan clock so that scan sequences can be inserted between vector sequences produced by the test generator, and presented a flip-flop design to ensure that no extra pin overhead above that for the normal scan design is incurred.

The above method does not depend on test pattern generation and is not expensive in computational terms, but it does not necessarily guarantee the same level of fault cover as full scan. In common with other partial scan methods it considers the circuit as a whole, without taking advantage of partitioning, functionality or architecture.

An extension of the Cheng method was described in [Gundlach90]. This presented a more detailed algorithm that considered not only cycles but also reconvergent fan-outs. It uses a graph model of the flat gate level description of the circuit to calculate a representative set of basic cycles and reconvergences, from which all cycles and reconvergences in the design can be created. A set of test cells is placed in such a way that all basic cycles and reconvergences are broken, by selecting the location that breaks the highest number of cycles and reconvergences at each step. A set of heuristics keeps the number of added test cells to a minimum, by identifying cycles and reconvergences that do not hamper testability (this happens when cycle nodes of reconvergency branches can easily be forced to 0 or 1), by the ordering of nodes which break the same number of critical paths, and by enforcing a threshold on the minimum number of paths broken, to avoid a superfluous model. The technique was tried in conjunction with Intelligen and showed a large improvement on the number of T-cells inserted as well as on the run times. The prediction of test points was much faster than the automatic test pattern generator (ATPG) process, enabling the ATPG to be run for a second time to improve the fault cover.

2.3.2 Built-in self-test systems

Built-in-self-test methods (BIST) are used either for off-line self-testing of the circuit, or for on-line checking of circuit functions [McCluskey85]. For example off-line BIST may use exhaustive test patterns or rely on a test program stored on the chip. On-line BIST systems may rely on pseudo-random methods, using pseudo-random signal generators (PRSGs) such as linear feedback shift registers (LFSRs) to generate the test signals [Williams83]. The test response is compressed into a signature by using single or multiple input signature registers [Smith80, Bhavsar86]. The aliasing error in the process determines the fault cover achievable. One test structure that can be used for both test signal generation and signature analysis is the built-in logic block observer (BILBO), a register that can be configured into latch mode, scan mode, PRSG mode, and multiple input signature register (MISR) mode [Koenemann79].

As in partial scan, BIST can be selectively or optimally applied to increase testability while minimising area overheads. An example of such an approach was Plessey's DEMIST (Design Methodology Incorporating Self-Test) [Ambler86]. The aim was to provide software tools for the automatic insertion of the ABIST method (Autonomous Built-In Self-Test), which is based on an enhanced BILBO register. Responsibility for test was to be effectively removed from the designer, in return for a number of design restrictions. The methodology provides the usual benefits of self-testable designs, namely a reduction in the need for expensive ATE, cost saving in test pattern generation and fault simulation effort (which are replaced by signature prediction), high fault cover, and testing at system speed.

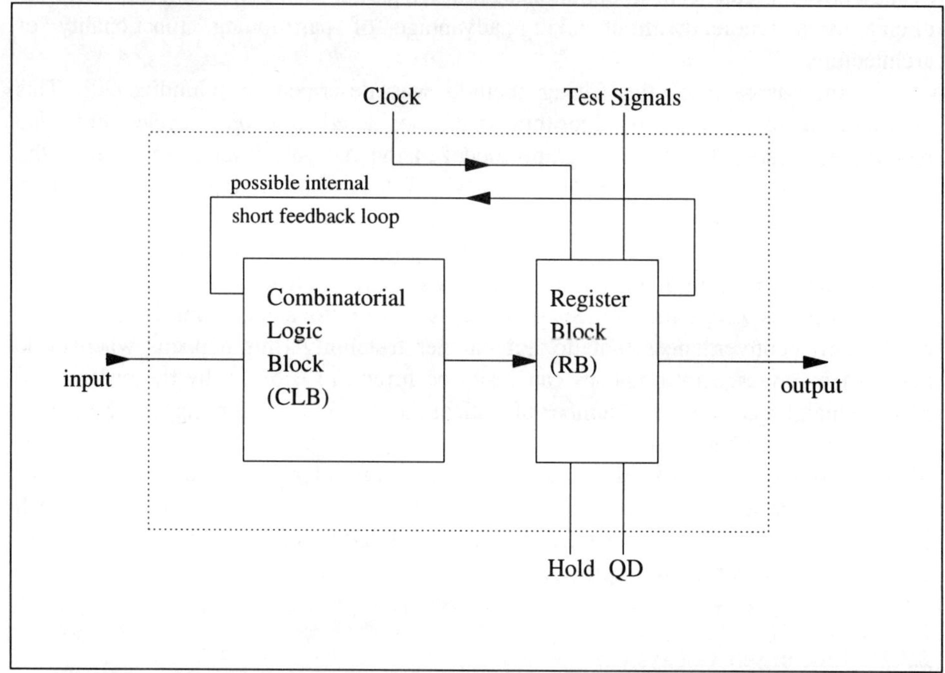

Figure 2.2 A structured/synchronous building block

The system is based on providing building blocks which would guarantee testability - as long as these blocks are used to realise the function of the circuit, ABIST could be successfully overlaid and therefore testability would be guaranteed. The building blocks are structured/synchronous building blocks with a restricted number of inputs to limit the length of exhaustive test sequences. Blocks are made up of a combinational logic block plus a register bank, as shown in figure 2.2.

The design enhancement for test is done by DEMIST. To facilitate self-test, output registers are replaced by structured test registers. These are BILBO registers with three extra states, of preset, hold and toggle. Blocks can then be linked in test chains, with one structured synchronous block generating test patterns while the second one analyses them. Multiplexers and test decoders are provided, in order to allow each block to cascade itself into more than one test chain. This is to take account of the fact that all blocks driving the block under test should be engaged in test pattern generation. The registers are therefore partitioned according to destination, before test chains are routed with the appropriate multiplexing and decoding circuitry assigned. The system then generates a test plan. Memory blocks and feedback loops not local to a single block can also be handled by this method.

This approach was shown to be economically viable, despite an area overhead of about 20%. This in itself is an interesting finding, showing that a technique which is relatively expensive in terms of area overhead can still be economically viable when all other factors are considered, and demonstrated the usefulness of economics modelling for test. The DEMIST technique itself is only suitable for synchronous designs such as signal and data processing applications, and not really applicable to inherently asynchronous designs such as switching networks. The preference for exhaustive test can lead to large test times, even at system speed, despite the fact that some chains can be tested in parallel.

This approach largely removes the responsibility for design for test from the designer in favour of the automated system. Although in the short term this results in testable designs with little designer retraining, the solution cannot be altered or optimised further by the user, and does not exploit any of the user's existing test knowledge. In the long term, and with different types of design considered, designer knowledge will have to be relied on, with automated DFT systems being used for advice and guidance.

Another system which automated the use of built-in self-test to improve the testability of a design was the Alvey BIST system [Jones86]. This and other knowledge based systems (KBS) for test strategy planning are described in more detail in section 2.4.1. Alvey BIST was a knowledge based test strategy planning system programmed in LOOPS, a Lisp based language. The system attempted to use weighted test resources in the design, with possible modifications, to create a BIST plan. Constraints of area overhead and test time are also considered, and backtracking is supported. The knowledge for the DFT planning was derived from a group of testability experts.

2.4 MULTIPLE STRATEGY SYSTEMS

Partial scan methods do have the advantage of reducing the area and test length drawbacks of full scan, but do not allow any true flexibility in DFT modifications to a design. The structure of the circuit is not necessarily fully exploited. In design terms, circuits are getting larger, but are often well partitioned into functional blocks. The use of advanced CAD tools encourages this process. In testability terms, a large number of structured test methods have been described, and these are often optimised for particular functional blocks. For instance, Zhu and Breuer [Zhu88] described over 20 different design for test methods for a PLA. This seems to point towards strategies which employ a variety of DFT methods, taking into account the structure of the circuit and matching test methods to circuit structures, often at an early stage in the design. These systems can be described as multiple strategy systems. It should be noted here that the boundary between single and multiple strategy systems depends on how broad the definition of a 'single' strategy is. For example, different implementations of scan path cells may or may not count as different DFT methods, depending on the broadness of the definition, as well as the requirements of the user.

A true multiple strategy system would not only consider a number of different test methods but also a number of different circuit structures, enabling real designs to be evaluated. Again, the use of advanced CAD tools makes designs easy to alter and is conducive to the implementation of a variety of DFT methods. The implementation of a design for test strategy may even be part of a synthesis tool [Fung86].

Multiple test strategy systems which attempt to allocate a variety of different test methods to a variety of testable units in an optimal way often need to create a description of the design from the test point of view. This was the case in the description of global design for testability (GDFT) nets, developed as part of the ESPRIT project CATE (Computer Aided Test for Europe) [Croft85]. The CATE system concentrated on partitioning the circuit as part of the planning process which allowed the designer to implement structures to allow test access to the testable units in the design. It aimed to provide a framework within which test strategy planning would be possible. The GDFT nets were described in text form, and included descriptions of switching functions, sequential depth, transparency, clocking schemes, as well as functional description of controllers. Test sources (such as LFSRs), test sinks (signature analysers), or structures that do both (e.g. BILBOs), were also considered. Representative test patterns could then be simulated. CATE aimed to provide a set of tools that made early DFT possible, rather than an automatic test strategy planning system.

Multiple test strategy systems may concentrate on allocating test resources such as LFSRs and signature analysers in an intelligent way, taking the structure of the circuit into account [Jones86]. These resources are then used to formulate test plans by using them to generate and observe tests. Examples are described in section 2.4.1.1.

Alternatively, structured methods for specific testable units can be coded and used in a functional block oriented approach which is especially suited to cell based designs, already conveniently partitioned into distinct blocks. An example of this approach is the Philips macro test [Beenker86, Beenker90]. Macro test was created in order to add testability to mainly digital signal processing chips, and was incorporated into the Piramid silicon compiler, (a silicon compiler for the synthesis of digital signal processors). It aims to achieve 100% fault cover (using fault models dedicated to functional blocks), and to reduce test pattern generation time by careful partitioning. Test techniques are selected for each macro cell in the design, suited to the macro and the chip architecture. The execution of each macro test is independent of its environment, hence the saving in test pattern generation. The individual macro tests are then used to form a full chip test, in a process known as test assembly. The testing of the system is controlled by a test control block (TCB), as shown in figure 2.3.

The macro test approach relies on design partitioning. The design is partitioned into testable macros, which are related to the functional building blocks in structured very large scale integration (VLSI) design. Macros must be independently testable, and accessible from the chip pins. These requirements are met by adding special circuitry, known as test interface elements. Macros should be of a uniform test type (e.g. RAM, random logic), and should not overlap each other. The partitioning itself is guided by several factors, among which are the limitations of the ATPG system (which may limit the maximum practical number of inputs to a block), the existence of fault models for specific structures (e.g. PLA crosspoint faults), the number of clocks in the system (partitioning ensures that macros are controlled by one clock, to avoid clock tuning problems), and the observability and controllability of nodes.

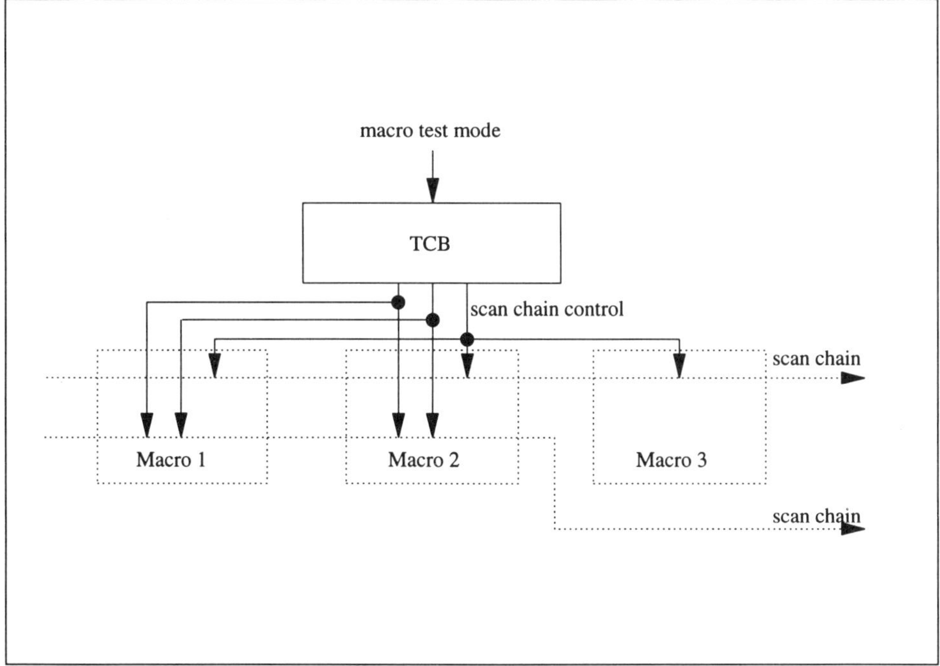

Figure 2.3 Macro cell leaves controlled by a test control block

In practice, when the macro test philosophy was implemented for Piramid, partitioning was based on implementing a set of design for testability rules. A fixed set of partitioning rules was produced, which exploited the design hierarchy. This was possible, as the silicon compiler dealt with a comparatively narrow set of designs, allowing a set of partitioning rules to be defined unambiguously. The designer can choose from macro specific test methods for each macro leaf. scan, BIST, partial scan, and test control logic are all supported by the silicon compiler, and the authors quote a 10% area overhead for 100% fault cover, with very short test generation times, as tests are generated independently for each macro. A test assembly process can then create the full test plan.

The Philips macro test is a successful system, which is currently in industrial use. It addresses specific types of design, using a small set of test methods and making it possible to have well defined partitioning and testability rules. This results in a very effective implementation of a test strategy tool, but also makes for an inflexible system. Whether this is a drawback or not depends on the user. Where designs can be classified into well defined sets, it can work very well. However, the designer has little guidance in choosing testability methods, a fact which the authors have pointed out.

An automatic multiple test strategy system often tries to perform the actions of a human testability expert. As a result, it is likely that knowledge bases and rule bases would be used, thereby removing the need for the designer to have up to date knowledge of every aspect of design for test structures.

2.4.1 Knowledge bases and test strategy planning

The problem of test strategy planning is a complex one, involving a variety of issues. The test methods chosen not only need to fulfil the technical specifications of compatibility with the design and each other and achievement of a specified fault cover, but also should take into account factors such as the test pattern generation systems available, limits on gate count, pin count, test pattern size, the ATE available, and the experience of the designers in implementing the testability methods. Some of these are conflicting requirements. For example, having an unlimited number of gates makes the task of achieving the required fault cover much easier than if a tight limit were imposed, but may increase area overhead. Such conflicts need resolving and decisions need to be taken at a local as well as a global level, the local level being circuit partitions and the global level being the overall design.

In test strategy planning, therefore, there are a variety of trade-offs to consider, which demand a high degree of expertise on the subject. Design for test relies on human experts for successful implementation. A human expert would have knowledge of different methods of design for test, their specific requirements, optimum implementations, and would also use a variety of heuristics to help the process. We have already seen that within a narrow scope such as partial scan path, well-defined methods exist to automate the test planning process. However, if an automatic system were to be successful in a variety of applications, it would need to store and use knowledge in a similar way to a human expert. Therefore, knowledge based (expert) systems, with the formalisation of knowledge they employ, are well placed to assist in the solution of this problem [Breuer88].

Currently a variety of knowledge based systems for test strategy planning have been described, with different levels of 'expertise', ranging from full rule based expert systems with explanation capabilities, to systems which use a knowledge base with a mainly algorithmic approach to inferencing. The broadness of application also varies, from systems which consider a single design type (e.g. PLAs [Breuer85]) to programs that can handle complete designs made up of many different types of functional block, such as TDES [Abadir85] and the EVEREST test strategy planner, which is described in detail later in this book.

There are two major requirements that knowledge based systems (KBS) have to fulfil. First, the knowledge to be used and the way it is to be formalised and stored have to be defined. Knowledge elicitation, the acquisition of knowledge from experts, is a notoriously difficult task, not only due to the large number of facts, but also due to the large number of links, rules and tricks which an expert would use, but would not necessarily be able to formalise and describe. It is therefore vital to describe the scope and aims of the system, and as far as possible the exact information needed to create the knowledge base. The better the definitions at this stage, the easier the knowledge elicitation process will be.

Closely linked to the stored knowledge is the inference process used, i.e. the set of rules and procedures that operate on the stored knowledge to solve a given problem. The inference engine can be a part of the knowledge base itself, in the cases where rules and data are stored together, or it is equally feasible for it to be a different entity from the knowledge base itself.

As knowledge based systems have been developed through research into artificial intelligence, languages such as Lisp and PROLOG, which are within the artificial intelligence domain, are frequently used for their implementation, as they have constructs that simplify coding KBS. However, there is often good reason why an algorithmic language such as Pascal or C should be used, depending on the specification of the system. This may be the case where the inferencing is mainly algorithmic, where mathematical calculations are involved or where speed is important. With an algorithmic language, any backtracking needs to be defined explicitly, which may make the development process more difficult, but as a result, backtracking becomes highly controllable and predictable.

The following section will describe a selection of knowledge based systems for test strategy planning, examining their philosophy, aims, type of knowledge and inferencing mechanisms, and, of course, their advantages and drawbacks. Probably the best-known work in this field is that of the University of Southern California with the TDES system, which will be discussed first.

2.4.1.1 TDES/TIGER and associated knowledge based systems

A knowledge based system to aid in creating easily testable VLSI chips, called TDES (Testable Design Expert System), was developed at the University of Southern California [Abadir85]. It was implemented in Lisp and designed to fit within the Advanced Design Automation (ADAM) system developed at the same university.

The aim of TDES was to create 'a framework for a methodology incorporating structural, behavioral, qualitative and quantitative aspects of known DFT techniques' [Abadir85]. The designer may then adopt a systematic approach to DFT provision in a circuit. The following is a brief description of TDES.

TDES operates on a register transfer level description of the circuit. The circuit has to be partitioned into kernels (testable units). Testable design methodologies (TDMs), held in the knowledge base, can be embedded within kernels. There are three basic structures recognised: combinational logic, registers and RAMs. Structures can be made up of the interconnection of basic structures. It is recognised that different implementations of basic structures may exist; for instance, combinational logic may be implemented as PLAs, ROMs, or random logic. The circuit description is stored as a hierarchical graph model. Kernels are structures with well-defined characteristics, for which at least one TDM is known. The circuit partitioning process divides the structures of the circuit into kernels, using a selection of simple rules, such as 'Any combinational kernel must have a single well-defined design style'. For instance, if a combinational structure was implemented as a PLA and a ROM, it would be treated as two separate kernels. Kernels may correspond to functional blocks such as control and memory, and are ordered for TDM embedding using a set of weightings, related to their type (some kernels such as registers are easy to test), controllability and observability, criticality (related to the choice of TDMs), size and regularity.

TDM data are stored in a frame based representation, which consists primarily of a structural template identifying the type, size, and style of kernel suitable for the methodology. Essential built-in test structures are described, as well as connection paths. An example template for the BILBO TDM is shown in figure 2.4.

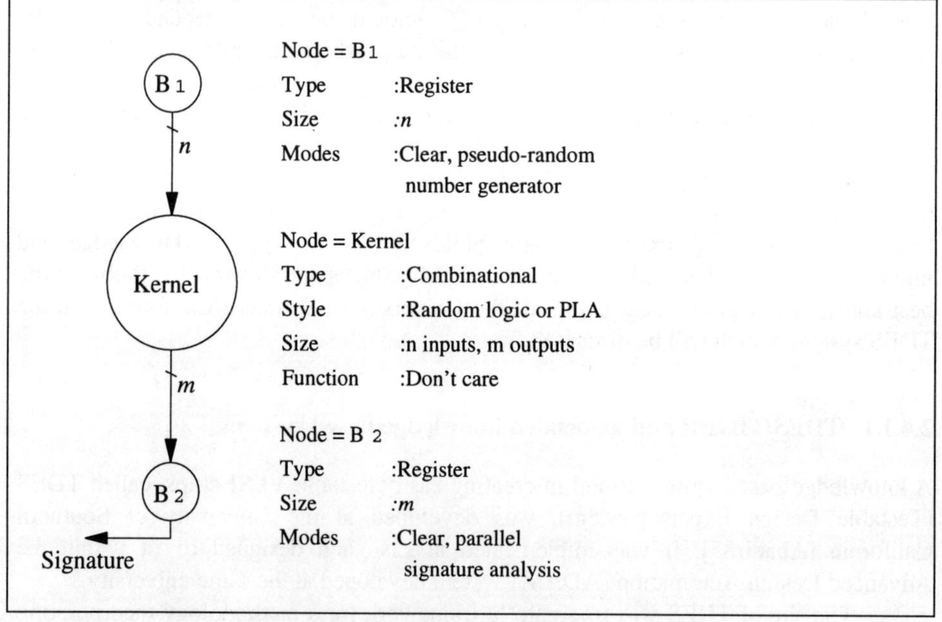

Figure 2.4 The BILBO TDM structural template

A test plan is also provided, which consists of the on-chip actions that constitute the life cycle of a single vector. Each step in the test plan is a group of actions that can be executed in a single clock cycle. There is also a set of TDM measures, which is intended to be used in the evaluation of TDM embeddings. Measures include area overhead, ATE requirement, TPG requirement, i/o requirements (extra pins), performance impact, test time and fault cover. This is intended to be used in the calculation of a cost function, using relative weightings specified by the user. There is also a measure of the ability to share built-in test structures between different kernels, which may influence the acceptance of an embedding. The test plan and measures for the BILBO TDM frame are shown in figure 2.5.

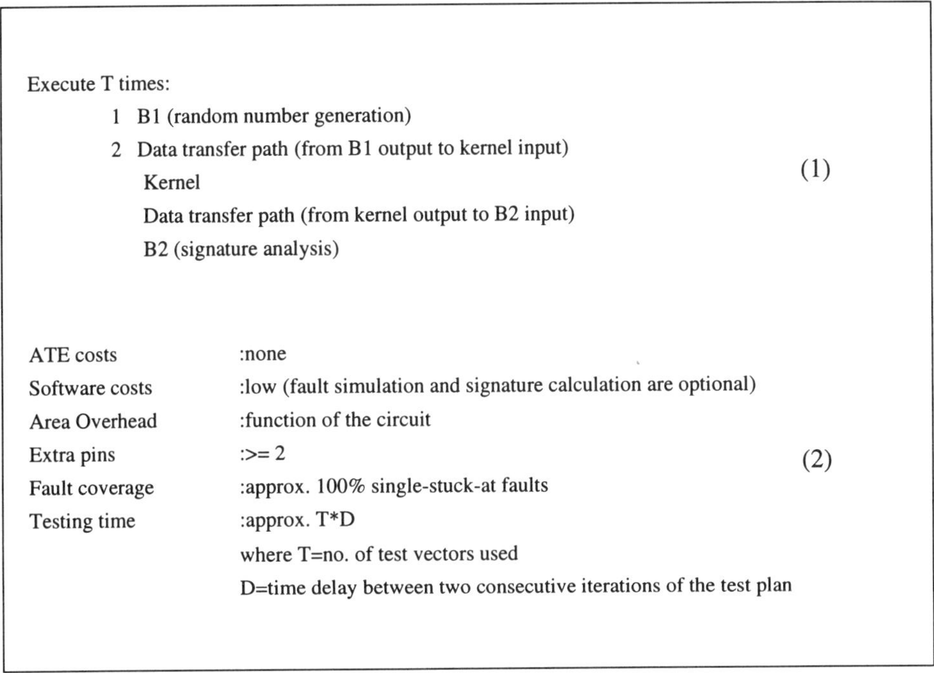

Figure 2.5 BILBO TDM. (1) The test plan. (2) Simple measures

The implementation of a TDM into a kernel is called an embedding. In order for an embedding to be performed, every structure in a TDM template needs to be matched with a structure in the circuit, and every connection between structures needs to be matched with an I-path between corresponding circuit structures. The concept of I-paths is important in the formulation of test plans. An I-path is an identity path, which means that it is a path through a kernel for which data can be transferred unchanged, after the fulfilment of any activation conditions. I-paths are calculated according to a set of rules before any embeddings are considered. Once all feasible embeddings have been generated for a kernel, measures for TDMs are calculated to determine the most suitable embedding. Note that the addition or modification of structures may be necessary to realise certain embeddings. A simple scoring function is used to compare embeddings, based on the TDM measures and user specified weights. This method is simple to implement and allows speedy calculation, but there is no guarantee that user specified weights are objective or that the TDM measure estimates are accurate. As a result, the user would need to closely examine the embedding comparison process in order to make sure that alternative, perhaps more suitable embeddings are not discounted.

Kernels are processed in the order determined at the beginning of the process. Each modification is taken into account for subsequent embeddings, and a test plan is generated. The circuit description is modified to implement the embedding. However, apart from estimates of the shareability of test resources and estimates of the effectiveness of a TDM in testing circuit structures other than the kernel under consideration, there is little global optimisation, but decisions are taken at a local (kernel) level.

The successor to TDES is TIGER (Testability Insertion Guidance ExpeRt [Abadir89]), which is now an industrial tool. It handles gate level descriptions, and uses a more sophisticated partitioning process than TDES. It uses a similar weighting system to evaluate embeddings, mainly in terms of area overhead, estimated TPG effort, and estimated test time.

Also developed at the University of Southern California, PLA-TSS [Breuer85] is an expert system specifically for testable PLA design. In this case TDM attributes are a function of PLA size and test requirements. The system attempts to find the test methods which are closest to the user's constraints and requirements by using penalty/credit functions for each attribute and combining these into a score in a linear fashion. Backtracking is possible when the system fails due to over-constrained requirements, in which case an analysis is performed to identify the most critical reasons for failure.

The Alvey BIST system [Jones86] was a test strategy planning system programmed in LOOPS, a Lisp based language. The work has strong analogies with TDES, but is exclusively directed towards BIST systems. The system examined test needs and attempted to use internal test resources in the circuit such as shift registers and BILBOs, which are either in existence or can be easily added. Each test resource is given a weighting depending on the number of blocks that can use it, and test resources with highest weighting are used first, much like the weighting system used in TDES. Constraints of area overhead and test time are also considered, and backtracking is supported. Test plans are generated by the propagation of test tokens through the circuit, noting resources found and transforming existing registers. An extension to the TDES I-path concept was introduced, which matched the transparency mode of the block under consideration to the type of test token that could be propagated through it unchanged. The concept was termed translucency. For example, if a pseudo-random test sequence could be propagated through a circuit block so that the output was also a pseudo-random sequence (although not necessarily the same), the block was said to be translucent to pseudo-random tokens. The Alvey BIST system simplified the test strategy planning problem by examining only BIST test methods. It concentrated on the task of providing a test plan, while optimisation attempts were confined to an estimate of the shareability of resources and keeping within designer defined constraints.

A system called DEFT [Samad86] provides testability advice similar to TDES. However, it concentrates on providing extensive explanations for any decisions taken. In order to do that, knowledge on top of that normally required for DFT insertion has to be included. The input is a circuit description with associated user defined testability objectives, such as maximum hardware overhead and minimum fault cover. The system selects testability schemes and modifies the circuit, allowing the user to question the reasons for particular actions and providing explanations. The knowledge base contains representations of general testability principles as well as specific DFT schemes. Knowledge takes three different forms: heuristic knowledge provides sets of actions to be performed given a set of circumstances, procedural knowledge defines a sequence of actions, and structural knowledge describes circuit elements. As in other systems, it is up to the user to specify the relative importance of test objectives and constraints, putting the objectivity of the DFT selection process in question. DEFT, like other systems, concentrates on providing a correct, rather than an optimal scheme.

Other knowledge based systems for test advice have different aims and approaches to TDES. Design for testability methods can be made part of the design synthesis route, and this is what the Silc automatic design for testability system aims for [Fung86]. The system, named TESTPERT, links design and testability issues in the silicon compilation process. It adopts a modular DFT philosophy that allows a mix of different test methods, although in practice it seems primarily scan oriented. It employs a testable by construction approach, and uses a test controller to coordinate testing and a testability bus to provide accessibility and is aimed primarily at the design of telecommunications chips. The main components of the system are the testability rules to enforce DFT, the testability evaluator, which operates on different levels, and the testability expert (TESTPERT) which chooses candidates for DFT.

Portions of the circuit which are difficult to test are recognised at several levels. The system can describe the circuit in terms of finite state machines, and a measure based on information flow can give a guide to testability problems. Additionally, the system employs an extension of SCOAP [Goldstein80] to calculate controllability and observability measures, as well as a path tracing algorithm that can be used later to provide guidance for partial scan placement. The testability expert then selects changes based on design constraints, the ATPG algorithm used, the test environment and 'cost' criteria supplied by the user. The chip data structure is changed mainly by manipulating finite state machine descriptions before the final parsing and minimisation are done, in order to take full advantage of the synthesis process. Testability by construction is achieved by adding logic to the finite state machine level, or using specially synthesised components with in-built testability which are coded together with their cost/benefit attributes (e.g. PLAs with BIST) in place of non-DFT components. The choice of DFT methods is limited, and is mainly scan based apart from pre-synthesised cells. Again, by the nature of the system the emphasis is on correctness rather than the ability to explore alternative solutions. A system called CATREE (for Computer Aided Trees) based on a binary tree representation of the circuit [Gebotys88], attempts to use test cost, silicon area and delay estimates as feedback to the synthesis process so that more testable designs can be found. However, there is no attempt to control the objectivity of these measures.

2.5 ECONOMICS BASED TEST STRATEGY PLANNING

The previous sections described a variety of systems which offer some form of test strategy planning. Some systems only offer variants of the same test method (single strategy systems), while others allow the user to choose from a variety of strategies (multiple strategy systems). However, the methods used for the evaluation of alternative strategies are often a weak point. For example, a simple weighting system based on user defined weights, as used in TDES [Abadir85], has no guarantee of objectivity and can miss a good solution. Attempting to minimise a single test related parameter such as area overhead is also not an objective measure, as the relative importance of the parameter in overall product cost as well as the value of trade-offs in a good solution are not considered.

2.5.1 Economic test strategy planning systems

All the above systems rely on subjective assessments of the relative importance of test and design related attributes. An objective evaluation method that would enable a comparison of test strategies, without the need for the user to be able to assess their relative importance, would be desirable. Test strategy planning decisions based on an economic analysis of the design and the economic effects of design for test methods are an objective way of comparing alternatives.

In comparing different DFT methods, a large number of parameters need to be taken into account which may not be directly comparable. An example is chip area (often increased by DFT methods) and test generation effort and test time, which may be reduced. The relative importance of these parameters is not clear and depends on the priorities of the person making the evaluation. For example, for a designer who is not concerned with test, the overriding concern would be the minimisation of area and maximisation of performance, while a test engineer would probably prefer to sacrifice some area and performance in return for easier testing.

In order to avoid a subjective evaluation, it is important that different parameters can be compared to each other. This can be done objectively by translating them all into the common measure of economic effects to the company, using cost modelling techniques. User defined constraints can also be evaluated using economics methods, in order to achieve general strategy planning. This is because user defined constraints (apart from these imposed by the product specification) should not be imposed without economic justification.

A set of related systems were developed on this basis. The first system was called BEAST (for Brunel Economic Aids for Strategic Test), and was developed under the Alvey programme [Dear89]. It was aimed at standard cell designs and used a knowledge base of test methods, programmed in PROLOG. The economics model was developed by the author, and was programmed in C. The modelling was deliberately kept very general, so that it could be tailored to different users' requirements. It provided a useful prototype which demonstrated the viability of the approach and led to the development of the EVEREST test strategy planner (ECOtest), in association with Siemens-Nixdorf under the ESPRIT project EVEREST [Dislis91]. This was a system aimed specifically at ASIC design, using a cost model tailored to the needs of the user company. It linked into Siemens design libraries and has been evaluated in industry. It was implemented entirely in C and, although it used the same elements of a knowledge base, economics model and test strategy planner as the BEAST system, it was much more specific and resulted in an industrial prototype. It fully calculated the economic and technical effects of test methods and also offered some design optimisation and a range of different test strategy planning algorithms, including simulated annealing, never before used for test strategy planning. The structure of ECOtest and the automatic test strategy planning algorithms are descibed in chapter 4.

Following the sucessfull completion of the ECOtest system the ECOvbs system was developed [Dislis91, Dislis93]. The system applied the same basic principles of economic analysis to board designs as ECOtest did to ASIC designs. It was designed to be used on its own or in conjunction with the ECOtest system to exploit hierachical usage of the tools. The structure of ECOvbs and economic analysis of board test is decribed in chapter 5.

2.6 SUMMARY

This chapter has described a number of systems that offer different forms of test strategy planning and has discussed the advantages and disadvantages of these systems. The test strategy planning approach that is the subject of this book was introduced, pointing out the advantages it offers over current systems. The next chapter describes economics modelling methods and gives a detailed account of the economics model the authors developed for the evaluation of test strategies for ASICs.

3

Economics modelling for test strategy planning

3.1 INTRODUCTION

This chapter will discuss economic modelling techniques in general, and proceed to describe economics modelling developed specifically for test. These techniques were used to develop the hierarchical life cycle test economics models used in the test strategy planning systems ECOtest and ECOvbs, which will serve as examples. The test economics model for integrated circuit testing will also be described as an example of how an economics model can be set up for a particular user company.

The terms 'economics model' and 'cost model' are often used interchangeably. However, we try to use the term 'economics model' in preference. The reason for this is that the term 'cost' is general. It is not only used for financial parameters but also for technical parameters, such as test time or chip area. Therefore the term 'cost model' is often used for models which calculate these non-financial parameters. Therefore one could define a cost based planning system that uses only one or two parameters on which decisions are based. The relationship between these parameters is defined by the economics model. For example, a cost metric based on area overhead and speed degradation would favour non-DFT solutions, on the basis that DFT increases area overhead and decreases speed. This is clearly not our approach, as we are looking for an objective basis for test decisions. In this book, cost is always related to financial parameters, and this is the reason why the term 'economics model' is used. An additional reason is the educational aspect. People mostly relate the term 'cost' to additional cost but not to cost savings. The point of using an economics model for test decisions is to arrive at a good solution which is also cost effective. The term 'economics' is much more related to cost savings or to the positive effect of a method or strategy. Therefore the term 'economics model' is better suited for its use in test strategy planning than the term 'cost modelling', because it automatically includes the saving aspect.

In the rest of this chapter, section 3.2 will describe economics modelling techniques by defining relevant terms, the various types of economics model, and the techniques for developing economics models. In section 3.3 methods for the estimation of life cycle cost will be described and discussed. Section 3.4 will present in detail the test economics model for ASICs, as an example of the development of a specific test related model, and section 3.5 will present a methodology for analysing the sensitivity of the model to the input parameters.

3.2 ECONOMICS MODELLING TECHNIQUES

The main objective of economics modelling is to provide a method for analysing the financial costs of a product or procedure or both before they actually occur. Economics modelling is a predictive technique for performing an economic analysis. An economics model allows the user to estimate the economics of a certain product by predicting the values of the cost-influencing parameters. A test economics model is an economics model, which is used to predict the economics of test strategies. The analysis is performed to make certain decisions, ranging from the purchase of equipment to methodologies such as design strategies. Therefore an economics model can be seen as a decision model [Dinkelbach69].

A *decision model* is built upon *parameters* and *variables*, which are used to determine a *target value* [Dinkelbach69]. In a decision model the target value is subject to optimisation by determining the optimal combination of the variables for a given set of parameters. The parameters are assumed to be fixed for a certain decision to be made, whereas the variables can vary within a defined range. Various techniques exist for determining the optimum mix of variables. These techniques are known as linear or nonlinear programming techniques [Luenberger73]. Using the above definition, the variables of the decision model are test strategy dependent, and the parameters of the decision model are test strategy independent. The linear programming techniques assume that a variable is continuous in its defined range. This is not the case for the test strategy planning task, because the variable values are fixed for a given test strategy, and each test strategy is defined by its own set of values for the variables. The problem of test strategy planning is one of combinational optimisation, and therefore different from the classical linear and nonlinear programming problem. Linear and nonlinear programming techniques are not suitable for the problem of test planning. Nevertheless the test economics model can be seen as a decision model, but with a different meaning of the term 'variable'. In order to avoid misunderstanding, the term *test dependent parameter* will be used instead of *variable*.

In order to consider all the effects of a decision, an economics model should consider all costs during the life cycle of the product, which is subject to the decision. Such a life cycle model considers all the costs from the design to the end of the useful life of the product. The process of economics modelling and its applications, such as test strategy planning, is composed of four objectives [Wübbenhorst84], and therefore it should be performed in four steps.

- The *interpretation objective* aims at recognising the costing relations and the structure of the cost. The cost structure is used to categorise the costing parameters in all phases of the life cycle. This structure can also be hierarchical, as this is the case in the test economics model.
- The *estimation objective* is aims in the development of costing relations, which describe the relation of the cost-affecting parameters. These relations can be modelled, for example, in mathematical equations.

- The *modelling objective* is related to modelling the cost for a given product by using the previous two objectives. This includes the gathering of input data for the economics model as well as the calculation of the cost.
- The *configuration objective* is the analysis of the calculated cost and the decision making process, which is based on this economic analysis. A life cycle cost analysis is defined in [Blanchard78] as follows: 'A life cycle cost analysis may be defined as the systematic analytical process of evaluating various alternative courses of action with the objective of choosing the best way to employ scarce resources'. This definition matches exactly the scope of test strategy planning. Therefore the configuration objective is the actual test strategy planning process.

The result of the interpretation objective and the estimation objective is the actual economics model. This economics model is developed once and is used multiply as part of the test strategy planning system. The modelling objective and the configuration objective are applied each time a test strategy planning session is performed. The test strategy planning systems ECOtest and ECOvbs provide features which support the process of achieving the last two objectives. The first two objectives are reached by the development of the economics model, which will be described in the following sections. The last two objectives can be reached by using the test strategy planning systems ECOtest and ECOvbs for a given electronic system, which provide features for gathering the data and calculating the cost (the modelling objective) as well as features for evaluating and assessing the various alternative test strategies.

3.3 METHODS FOR THE ESTIMATION OF LIFE CYCLE COST

The determination of the life cycle cost has to be performed very early in the design phase. This is because the decision on the test strategy includes design decisions which have to be taken before the electronic system is fully implemented. This leads to the major problem of economics based test strategy planning. The life cycle costs are not known in this phase. They must be predicted, and in fact this prediction is subject to uncertainty. This problem cannot be solved completely, but by using various methods the risk, which is related to this uncertainty, can be minimised. Some ways of handling uncertainty and risk are discussed towards the end of this chapter. Another method is the incorporation of the cost driving persons into the process of the modelling objective. This fact is discussed in several publications (e.g. [Hills85, Kirk79]).

Several methods are proposed in the literature in order to estimate costs [Madauss84]. Table 3.1 outlines and classifies these methods.

Judgmental methods are typically used in the very early phase of a product, where no detailed information about the product is available. These methods are based upon expert judgements, analogies and estimations of a rough of order of magnitude. The estimations can be driven by subjective criteria, and the application of these methods is very limited.

Because of this, many authors favour the use of *parametric methods*. These methods can lead to detailed results even in the early phase of a product's life. But their application requires a certain level of knowledge about the design, production and field phases and methods. A parametric cost model is a cost model which consists of primary and secondary parameters. The primary parameters are the cost-driving factors, and the secondary parameters are costs or cost factors, which are derived from the primary parameters or other secondary parameters by a mathematical equation. The most important parametric method is the cost estimation relationship (CER) method. It uses the experience from similar previous projects or products to develop a relation between cost driving factors and the costs. This allows the usage of statistical methods such as regression analysis.

Table 3.1 Cost estimation methods [Madauss84]

Methods	Application requirements	Areas of application	Limitations
Judgement • expert judgement • educated guess • rough order of magnitude	• expert's experience • rough product definition • analogous material	• early stage • situations without risks • independent • cross-checks • budget estimations	• subjective • undefined accuracy • not applicable for price negotiations
Parametric • cost estimation relationship (CER) • statistics • models • cost formulae	• historical data • regression analysis • CER material	• concept comparisons • budget planning • tender analysis • independent cross-checks	• extrapolation of data bases and models is often difficult • estimation accuracy can be vague
Detailed • work package estimation • work preparation estimation • cost formulae • engineering cost estimates	• time schedule • statement of work and specification • detailed technical material • price tenders	• situations with high risk • price negotiations	• expensive and time consuming • low flexibility • can lead to cost increases

Detailed methods are mostly based upon work packages. The project to be estimated is broken down into small work packages, for which an accurate estimation is possible. The detailed cost estimates assume much more information about the project than the parametric method, which may not be available when the estimation is performed.

A combination of detailed and parametric methods was selected for the development of the test economics model. The test economics model is structured into several smaller sub-models, which are related to work packages. This structure will be presented later in this chapter. The estimation of the costs of the work packages is performed either by parametric models or a mixture of a parametric method and a detailed method. For example, the design time of the ASICs is calculated by a parametric method. The calculation of the assembly cost of a board is based on the number of components per assembly type and the assembly cost per type. The calculation of the assembly cost is a detailed method, but the calculation of the assembly cost per assembly type is based on previous productions and is therefore parametric.

This mixture of the two estimation methods is a good trade-off between the accuracy of the prediction and the cost for the prediction process. The test economics model is intended to be used as the last step of the design specification. The structure of the test economics model reflects the detailing level of a specification. Most of the information about the parameters, which must be defined in order to determine the equations, is available at the end of the specification.

3.4 THE TEST ECONOMICS MODEL FOR ASIC COMPONENTS

The test economics for ASICs facilitates the prediction of all costs influenced by the test strategy. The model is designed to fit the development of cell based ASICs (semi-custom ASICs, which are developed by using a cell library). A cell library typically contains:

- simple cells like AND gates or inverters
- more complex cells like counters or shift registers
- macro cells like PLAs and RAMs.

The model is intended for use by ASIC designers and their managers. Test strategy planning should be done hierarchically in the same way as the design itself. Values for the parameters, which vary from design to design, should be easily accessible, e.g. through a CAD data base. Initial effort is required for data gathering to derive the design independent primary parameters. However, the cost for the day to day use of the model should be negligible. This was an important aspect of the model development.

The costs forming the model are divided into three parts:

- The *design costs* are the costs related to the design and development of cell-based ASICs.
- The *production costs* are the costs of the VLSI supplier. These costs are not calculated by a model, because the user of the model (the designer/manager) cannot influence the costs by controlling the production process. Nevertheless the price is influenced by characteristics of the design, especially the gate count. A data base of relevant parameters can form the customer-supplier interface.

- The *test costs* are the costs related to testing purposes. These include the costs for fault simulation and test pattern generation. ATE costs as part of the production are only considered if the ASIC price is influenced by the size of the test set. Costs for an incoming test are usually related to board cost.

3.4.1 The design costs

In the modelling of the design cost, the time to design (the main influencing factor) was linked to the complexity of the circuit. The design costs depend strongly on the designer's environment. Looking at real data about design schedules, it was found that the complexity of the design cannot only be measured in terms of gates. The correlation between design time and the number of gates is not very strong. In some cases the Parkinson principle ('work expands to fill the available volume') is the most suitable model for estimating design time. Nevertheless it should not be the aim of test economics modelling to follow this principle. The modelling was based on design effort for known designs, and calculated the engineering effort required for the project [Fey87] in order to arrive at design costs. The modelling considered the familiarity of the designer with the type of circuit under development, and also how effectively the CAD tools could be used for the particular design (the productivity of the design system). Design for test can affect the design effort if the method used is very new to the designer, not easily implemented from the CAD library, or alters the performance of the design in a critical fashion. These effects are estimated by the user (performance implications) and by information from the test method data base which links to the CAD library.

The model must be used with the following assumptions :

- The ASIC to be developed is a semi-custom design as described earlier.
- A top-down design by taking advantage of the hierarchical architecture is used.
- The CAD-system runs on a workstation. For workstations the CPU time costs are usually included in the engineer's cost rate. If an outside design centre is used or some of the CAD tools run on a mainframe, these additional costs are drawn explicitly.

The following design phases are taken into account for the prediction of design costs:

- initial development of the circuit
- design capture
- simulation and verification

The calculation of the design time is not only based on the gate count but also on other design parameters and design environment parameters. The design parameters are the gate count, the number of cells, the pin count, the originality of the design and the performance criticality. The design environment or productivity parameters are the productivity of CAD system which is used, the experience of the designers and the productivity of the cell library, i.e. how well the functions of the cells provided by the cell library match the actual design.

The parameters used in the calculation of the design time are as follows:

- The local complexity of each testable unit (TU) i is given by:

$$lcompl_i = cperf_i \times (1 - or_i) \times cgate_i$$

Here $cperf_i$ is the performance complexity for the TU. This is a linear factor used to relate the extra complexity due to DFT to the design time. A value of 1 indicates no change, a value of ∞ indicates that it is impossible to perform the design. The originality factor, or_i is a linear factor used to describe whether some of the architecture, functions or algorithms have been designed before, and in what way this knowledge accelerates the design process. The value of or_i varies between 0 (design is fully original) and 1 (the TU has already been designed). $Cgate_i$ is the gate count of the block.

- The pin complexity metric (*cpin*) is given by:

$$cpin = 2 \times in + out + 3 \times bi$$

where *in* is the number of input pins, *out* the number of output pins, and *bi* the number of bi-directional pins.

- The overall complexity (*ocompl*) is a function of the local complexity per functional block and the pin complexity, relating the number of pins:

$$ocompl = cpin \times \left(\sum_{i=1}^{blocks} lcompl_i \right)^{cexp}$$

where *cexp* is an exponent factor relating the design complexity to the gate count.

- The productivity metric is comprised of the productivity of the cell library, productivity of the CAD system and designer's experience.
- The productivity of the cell library (*plib*) is modelled by the following equation:

$$plib = \frac{\left[\dfrac{\sum_{1}^{i=blocks} fun_i}{cells + blocks} + \dfrac{2 \times \sum_{1}^{i=blocks} fun_i}{cells + fdblocks} \right]}{3}$$

where *fun* is the number of functions per TU, linking the functional to the test hierarchy, *cells* is the number of cells per chip, *blocks* is the number of TUs, and *fdblocks* is the number of functionally described blocks (in some cases, some blocks may not have a functional description associated with them).

- The productivity of the CAD system (*pcad*) is a linear parameter, ranging 0 to 1.
- The designer's experience (*pdes*) is modelled by

$$pdes = 1 - \frac{1}{kdes + exper}$$

where *exper* is the number of designs the designer has completed, and *kdes* is a normalising factor, normalising the productivity of a designer with no experience to that of an experienced designer.

- The overall productivity then, is expressed by

$$prod = kp \times pcad \times pdes \times plib$$

where *kp* is a proportionality factor.

- The manpower (*mp*), i.e. the engineering effort in terms of time is given by:

$$mp = \frac{ocompl}{prod}$$

- The design time (*destime*), i.e. the actual overall time taken, is a function of the manpower, fitted to historical data

$$destime = \mathrm{fn}(mp)$$

- The cost of using a mainframe (instead of workstations where the cost is incorporated into the designer's pay) is as follows:

$$mainframecost = cputime \times equrate$$

where *cputime* is the CPU time in hours that accounted equipment such as a mainframe is used, and *equrate* is the cost per CPU hour.

- The cost of using an external design centre (*descencost*) is given by

$$descencost = percuse \times destime \times descenrate$$

where *percuse* is the percentage of the effort undertaken by the design centre, and *descenrate* is the weekly cost of the design centre.

- The engineering cost (*engcost*) is given by:

$$engcost = mp \times costrate$$

where *costrate* is the weekly cost per designer

- The design cost (descost) is given by:

$$descost = engcost + descencost + mainframecost$$

Test pattern generation and fault simulation are also part of the design process, but in the test economics model this effort is part of the test cost.

3.4.2 The production costs

The production costs are modelled by the detailed method, and the risk of predicting those costs is on the ASIC supplier's side. Nevertheless the price depends on several parameters. The values of these must be known, so they are part of the design specification. The idea is to call for price offers for several sensible test plan configurations. Several ASIC suppliers confirmed that this would be possible. The production costs are also influenced by negotiations. The prices depend on particular market interests of the supplier. If, for example, a supplier wants to secure business from a company, they would probably start with low price offers. Also it is possible to have more gates on the chip than predicted but not to pay more.

To set a price for a specific ASIC the supplier must have knowledge about the following data :

- Production Volume (*vol*): Usually the volume is not one number. The expected delivery volume has to be quoted for every year. Typically the volumes over five years are predicted. For example, the prices can differ, if you have for the same ASIC a production volume of 100 000, and the delivery is distributed in one case over two years and in the other case over five years.
- Complexity: The complexity is usually quoted in 'number of equivalent gates'. The number is based on the equivalent gate number of the cells used. It usually includes the gate number of the i/o pads and excludes all macro cells. The complexity of the macro cells is determined by their parameters (e.g. the number of bits for RAMs).
- Pin number: The number of pins has influence as well on the die size as on the package size.
- Number of test patterns: Some suppliers limit the test set length. If the actual length exceeds this limit, the customer has to pay more. This limit depends on the pin memory size of the ATE.
- Technology: Technology of the silicon.
- Package type: Package size and stock.

The chip price is composed by two different cost parts. The non recurring engineering (*nre*) charges include all services from the supplier. The NRE charges are incurred per design. What they really include depends on the customer-vendor interface (see [Ardeman87]). The production unit cost (*puc*) is related to all costs incurred during the production of the chip. They include the costs for silicon, package, wafer production, packaging and testing. These costs are incurred per production unit. If there is a surcharge for long test sets, this additional price is part of the test costs. In the economics model, a pricing function is included, relating the production unit cost to the number of gates in the design. An example pricing function is shown in figure 3.1.

In the economics model, the production cost (*pc*) is expressed by:

$$pc = nre + vol \times puc$$

3.4.3 The test costs

The test pattern generation costs depend on the circuit, the fault coverage to be achieved, the performance of the test pattern generator and the type of test pattern generation (sequential or combinational). Deterministic test pattern generation can be performed by a tool (ATPG) or a person (manual test pattern generation, MTPG). The generation of random test pattern causes no test pattern generation cost. So the costs for test pattern generation for all self tests, which are based on random patterns, are negligible.

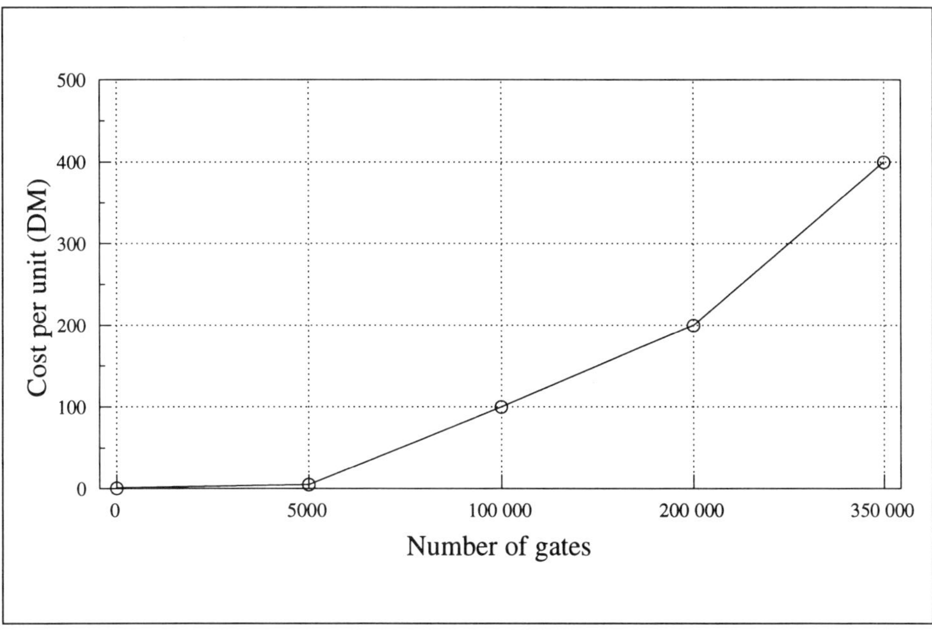

Figure 3.1 Production unit cost pricing function example

Manual test pattern generation is used where the ATPG system reaches its performance limits, or where an ATPG system is not available. Problems for ATPG systems usually occur with circuits of large sequential depth, and for asynchronous circuits. MTPG is used for faults undetected by the test pattern from ATPG and the functional patterns used for the design verification. It is also used for generating the propagation of patterns through the circuit. The model for ATPG cost is based on [Goel80]. The evaluation of ATPG data measured for scan based circuits showed that there is no significant correlation between the CPU time and circuit characteristics other than the gate count.

Most of the ATPG systems for sequential circuits multiply the circuit for every time frame. Therefore the effort depends on the average sequential depth of the circuits as well as the number of gates.

The correlation of test pattern generation costs and fault coverage is also derived from [Goel80]. After reaching a fault coverage between 80% and 90% very fast (exponential shape, phase I), the curve becomes linear (phase II). This linearity is probably caused by the fact that now most of the generated test pattern covers just a few additional faults. The slope depends on the fault coverage, where the slope becomes linear, and the complexity of the circuit for test pattern generation. The modelling does not include the effort of test pattern propagation (accessibility modelling), as full accessibility provision is one of the objectives of the test strategy planner.

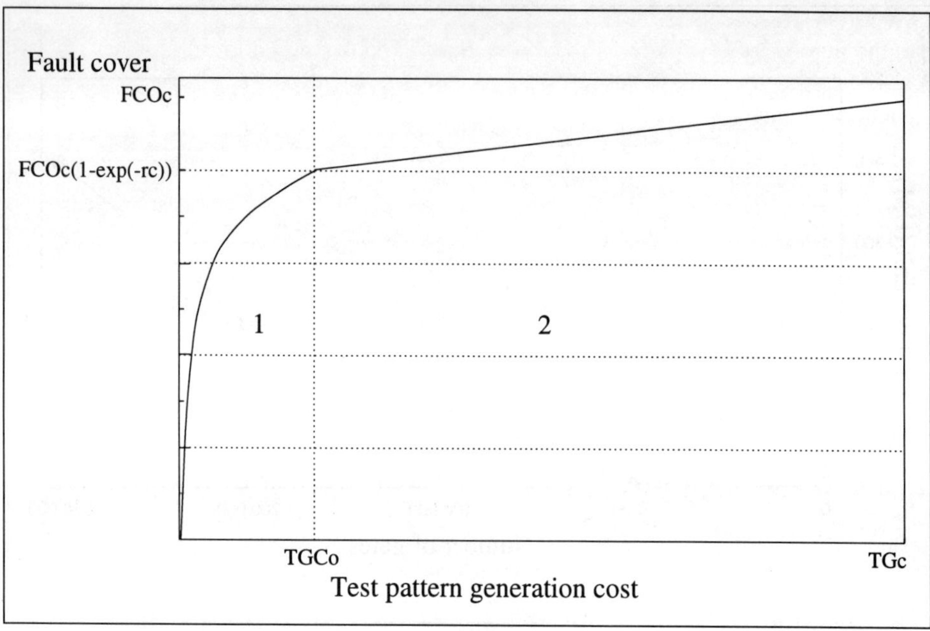

Figure 3.2 Test pattern generation modelling

The modelling equations are given below:

- The ATPG cost for the *i*th TU ($costtgs_i$) is given by:

$$costtgs_i = ktgs \times \left[cgate_i \times \left(avs_i + 1\right)\right]^{exptgs}$$

where *ktgs* is a linear normalisation factor, avs_i is the average sequential depth of TU *i*, and *exptgs* is an exponential factor linking the ATPG time to the gate count.

- The overall ATPG cost (*costtgs*) is simply

$$costtgs = \sum_{i=0}^{blocks} costtgs_i$$

The automatic test pattern generator is only run until it becomes impractical to continue, and the remaining faults, if any, are generated manually (a significantly more expensive process), corresponding to phase II of the graph.

- The fault cover achieved for TU *i* by the ATPG is calculated as follows:

$$fach_i = \left[1 - 2.72^{\left(\frac{-4.6 \times sdlim}{avs_i}\right)}\right] \times \left[1 - 2.72^{\left(\frac{-4.6 \times glim}{cgate_i}\right)}\right]$$

where *sdlim* and *glim* are the typical sequential depth and gate count respectively for which the fault cover of the automatically generated test patterns remains under 99%.

In order to calculate the manual test pattern generation cost, we need to estimate the number of test patterns to be generated. This we relate to the number of faults.

- The number of stuck-at faults of the *i*th TU is

$$faults_i = cgate_i \times fpg$$

where *fpg* is the average number of stuck-at faults per gate. The total number of faults (*faults*) is the sum of $faults_i$. If the achievable fault cover ($fcach_i$) is lower than the required fault cover ($fcreq_i$), then MTPG is necessary, otherwise ATPG is enough.

- The remaining number of faults ($remfaults_i$) for the *i*th TU (assuming $fcach_i < fcreq_i$) is given by

$$remfaults_i = \left(fcreq_i - fcach_i\right) \times faults_i$$

- The total time required for MTPG (in weeks) is modelled by

$$mtgtime = \sum_{i=1}^{blocks} \frac{remfaults_i \times tpf_i}{hpw}$$

where tpf_i is the average number of test patterns per fault needed for TU *i*, and *hpw* is the number of working hours per week.

- The cost of MTPG (*mtc*) is:

$$mtc = mtgtime \times costrate$$

where *costrate* is the weekly cost per designer/engineer.

In modelling the test application costs, it was assumed that test patterns are supplied to the vendor and are applied by the vendor's ATE, thus removing the need to model ATE running costs explicitly. The price is dependent on the total number of test vectors, including verification vectors. The number of test patterns for a TU is calculated based on an estimate of the average number of test vectors per fault, or is retrieved from the test method data base. However, as at present the system does not calculate the optimum scan chain configuration, the increase in the test vectors due to scan cannot be calculated reliably. If scan methods are used, the availability of a scan ATE is assumed, and scan test vectors can be identified as such by the system.

Typically a number of test patterns are 'free', i.e. included in the fabrication price. Test vectors that exceed that number are charged separately according to a pricing function. Typically, this is related to the memory size of the ATE pins. A charge is made each time a test vector set is loaded. For example, if each pin has 32K of memory, then the first 32K of test vectors would be applied free. Then a fixed charge per test pattern is made for each extra 32K set. An example pricing function is shown in figure 3.3.

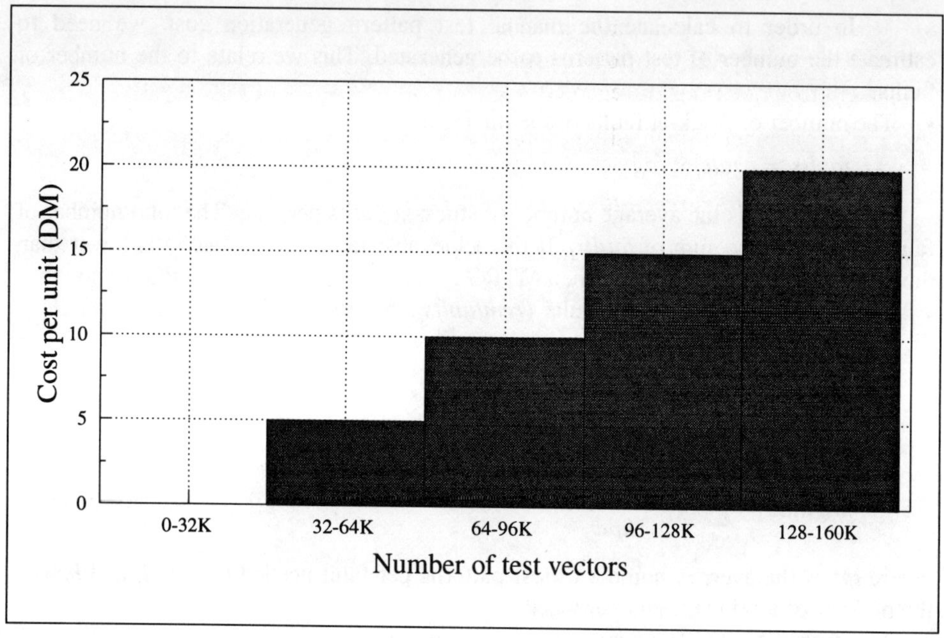

Figure 3.3 Test application costs pricing function example

The test application costs are dependent on the number of test patterns. The number of test patterns needs to be estimated. In some cases, such as RAM testing where the test set is derived algorithmically, the number of vectors is known and placed directly in the economics model. In more general cases, the following calculations are used:

- The total test set length (*tsl*) is the sum of the vectors generated automatically (*numtpa*), the vectors generated manually (*numtpm*), and the verification vectors (*tpver*).

$$tsl = \sum_{i=1}^{blocks} numtpa_i + \sum_{i=1}^{blocks} numtpm_i + tpver$$

where *i* denotes the *i*th TU.

- The number of automatically generated test vectors for TU *i* is:

$$numtpa_i = faultsa_i \times tppf \times (avs_i + 1)$$

where *tppf* is the test patterns per faut and $faultsa_i$ denotes the number of faults per TU for which test patterns were generated by the ATPG system, and is given by

$$faultsa_i = (fcach_i - fcv) \times faults_i$$

where *fcv* is the typical fault cover achieved by verification patterns. If the achiveable fault cover by ATPG is lower than that achieved by verification patterns, the test set length from APTG is disregarded.

- The number of manually generated test vectors for the *i*th TU is given by:

$$numtpm_i = remfaults_i \times (avs_i + 1)$$

- The test application cost (*tac*) is given by:

$$tac = \frac{(tsl - \text{remainder\_of}(tsl \div pms))}{pms} \times pps \times vol$$

where *pms* is the interval in the test application pricing function, and *pps* is the price per additional test pattern set per device.

- The total test cost (*tcost*) is given by:

$$tcost = tac + mtc + costtgs$$

3.4.4 Time-to-market considerations

Time to market is an important factor in the success of a product, in terms of both the sales window and the market share. However, costs relating to time-to-market effects are difficult to model as they depend on the current competition and long term marketing considerations which are beyond the scope of a test strategy planning economics model. Therefore time-to-market costs were not modelled in any detail.

However, many supplier-customer contracts contain clauses for penalties incurred for non-delivery and for the product not performing to specification. This kind of clause can be modelled fairly easily and is included in the economics model. Estimates of non-delivered units and performance levels have to be supplied by the user. Although not really useful for planning marketing strategies, this modelling can give an indication of the importance of considering time-to-market costs.

- The time-to-market cost (*ttmcost*) is modelled by:

$$ttmcost = npcost + plcost$$

where *npcost* denotes the cost of a non-performance penalty, and *plcost* is the profit loss due to non-delivered units caused by schedule slippage.

- The profit loss is calculated by:

$$plcost = ppu \times ndu$$

where *ndu* is the number of non-delivered units and *ppu* is the profit per unit.

- The total cost is the sum of the design, production, test and time to market costs.

3.5 SENSITIVITY ANALYSIS

One of the main drawbacks associated with the use of economics models is the risk of arriving at misleading decisions as a result of inaccuracy in the model's input parameters. A carefully constructed economics model would take into account the difficulties in data collection by, for instance, not requesting parameters which the user would have no way of accessing. However, as the test strategy planning process takes place early in the design cycle, many parameter values are estimates, and, as a result, subject to inaccuracy. This is partly due to the trade-off between the accuracy of the data and the effort or cost of the data gathering task. It is therefore important in the context of economics modelling to study the impact of the accuracy of the input parameters to the resulting cost prediction. There are two main reasons for this: first, to quantify the confidence of the economics model predictions; and second, to estimate the effects of the trade-off between data gathering effort and the accuracy of the results.

In previous work ([Dislis91, Varma84, Dinkelbach69, Myers83, Davis82]) sensitivity analysis was always performed by varying one parameter and keeping all other parameters constant. However, this method does not show the real impact of inaccuracies in the input data. The reason is that the sensitivity of the parameter under consideration is calculated statically, i.e. with static values for all other parameters. This creates a problem, as there is an implicit assumption here that only one parameter at any one time is subject to inaccuracy. It may be the case that certain parameter *combinations* create maximum sensitivity in the model results. This is an aspect that cannot be examined by static sensitivity analysis. Therefore, the assumption that only one parameter varies at any one time is not realistic, and a different evaluation method is needed which takes into account combinations of input parameters in order to evaluate the sensitivity of the output. This type of sensitivity analysis is called *dynamic sensitivity analysis*, because the sensitivity calculation is performed by dynamically varying all other parameters.

A dynamic sensitivity analysis study for the ECOtest model was undertaken, using Monte Carlo simulation, resulting in three sensitivity analysis applications:

- A *general sensitivity analysis* to classify each parameter of the cost model with regard to its sensitivity to the cost in general, i.e. independent of a specific case.
- An *iterative sensitivity analysis* to perform a detailed examination of the input data of the cost model iteratively. This is designed to find a set of parameters whose accuracy is critical, i.e. will always influence the accuracy of the total cost.
- A *total variation sensitivity analysis* to study the probability density function of the total cost, which is based upon the inaccuracy of the estimated input values, where the inaccuracy is handled as a variate with a known distribution function.

3.5.1 Monte Carlo methods for sensitivity analysis

The inaccuracy of the input parameters can be seen as a variate with a defined distribution. The type of the distribution is a uniform distribution, if the estimate defines a range, or a limited normal distribution, if the estimate is given by a mean value and a deviation from that mean value. The limited normal distribution is used to account for the fact that certain parameters, such as gate count, have defined limits (e.g. gate count cannot be smaller than zero).

The sensitivity analysis problem can be broken down as follows:

1. Determine the mean value of the final cost and its variance by solving the integral of the economics model in order to determine the sensitivity of the resulting cost with regard to the inaccuracy of the input values, and the sensitivity behaviour of the resulting cost with regard to the variation of one parameter with inaccuracy of the other parameters.
2. Solve the differential of the economics model in order determine the maximum sensitivity of the resulting cost with regard to the variation of one parameter with constraints for the other parameters. (This is an optimisation problem, which can be solved by an extrema analysis.)

Problem 1 can be solved analytically, but this would require the solution of multiple integrals over a very complex, discontinuous function, i.e. the economics model. Problem 2 cannot be solved analytically without modifications, which are related to discontinuities in the economics model, because they cannot be differentiated. Although the differentiation of the economics model could be performed analytically within the continuous regions, this approach is not practical. The analytical method requires the solution of very complex integrals and differentials, strongly tied to the related function. This function changes with the test strategy, as the mix of test methods varies. This would imply that the integration/differentiation would have to take place for each different test strategy, making the complexity of the problem (an the likelihood of errors) such that it would be unrealistic to pursue an analytical solution.

For these reasons, the Monte Carlo method was chosen for solving the integration problem and the optimisation problem, for the following reasons:

- The method is independent of the function to be analysed and therefore it is very flexible in its application to changing functions.
- The input parameter values are variates. Monte Carlo simulation is also based on variates [Rubinstein86], and therefore the method fits the nature of the underlying problem very well.
- The method can be used for both the integration and the optimisation problem [Hammersley65, Rubinstein86].

The essential feature of Monte Carlo simulations is that a random variable is replaced by a corresponding set of actual values, having the statistical properties of the random variable [Hammersley65]. This random variable can be part of the process under consideration, or it can be modelled from a deterministic variable. The actual set of values is used to analyse the process. This procedure is called Monte Carlo simulation.

'Problems handled by Monte Carlo are of two types called probabilistic or deterministic according to whether or not they are directly concerned with the behaviour and outcome of random processes' [Hammersley65]. The type of Monte Carlo used here is deterministic, because the process, which is simulated by the economics model, is deterministic. The behaviour of the economics model is not random, and the behaviour of the input data is also known. Also, in theory, the problem can be solved deterministically, as shown above. Hammersley defines deterministic Monte Carlo as a numerical solution of a deterministic problem. Deterministic Monte Carlo simulation is also called *sophisticated Monte Carlo*.

Consider the following example. A process is defined by the economics model and some input data, which are normally distributed. The question to answer is, 'What is the probability that the resulting cost will exceed a certain limit?'. This question can be answered through Monte Carlo simulation as follows:

- Generate random numbers for the normal distributed input parameters having the statistical properties of the given normal distribution.
- Calculate the related resulting cost value.
- Repeat the above procedures many times.
- Measure the probability of exceeding the limit by calculating the relation of resulting cost values exceeding the limit to the number of calculations.

Instead of solving the complex multiple integral of the economics model over the distribution function of the input parameters, we simply measure the result by performing Monte Carlo simulation, using random numbers with given distribution properties.

3.5.2 General sensitivity analysis

The general sensitivity analysis was a study of the economics model in order to classify all parameters with regard to their sensitivity impact on the total cost. This analysis gives a general idea on which parameters must be estimated very accurately (even if the effort involved in this estimation is high), and for which parameters a rough estimate may fulfil the accuracy requirements. An outcome of this study may even be that some of the parameters can be neglected for the cost evaluation. This fact would allow a simplification of the cost model by cutting out the effect of these parameters. The refined cost model would provide the same results with lower costs in data acquisition and test strategy planning.

The sensitivity classification of the parameters was performed by estimating for each parameter:

1. The mean value and the variance of the sensitivity in a constrained space $D \in R$ of the input parameters.
2. The maximum sensitivity of the total cost in a constrained space $D \in R$ of the input parameters.

As parts of the cost model are test method dependent, the sensitivity analysis was performed for three different representative test strategies. These are *no DFT, scan path* and *circular self-test path* [Krasniewski88]. The sensitivity value is defined here as the relative difference in the total cost of a relative difference of the analysed input parameter (i.e. calculating the effect on the output of a variation of the inputs). Table 3.2 shows the distribution characteristics of the cost model parameters used.

Table 3.2 Distribution characteristics of cost model parameters

Parameter name	Abbreviation	Lower limit	Upper limit
Number of cells	*cells*	3	64 000
Complexity exponent	*cexp*	0.8	1.0
Number of gates	*cgate*	1000	100 000
Labour cost rate	*costrate*	$500	$5000
Performance complexity	*cperf*	1	3
CPU time	*cputime*	0	1000 h
Design centre cost rate	*descentrate*	$10 000	$40 000
Computer equipment rate	*equrate*	$25	$1000
Designer's experience	*exper*	0	100
Required fault coverage	*fcreq*	70%	100%
Average number of faults per gate	*fpg*	2	5
Constant factor concerning designer's productivity	*kdes*	1	2
Constant factor concerning total productivity	*kp*	70 000 000	90 000 000
Manual test generation time per fault	*mtgtime*	0.05 h	1 h
Design originality	*or*	0	1
Productivity of the CAD system	*pcad*	1	5
Percentage of design time an external design centre is used	*percuse*	0%	100%
Number test pattern for which test application cost increases	*pms*	64 000	640 000
Test application cost per step	*pps*	0	$25
Production unit cost per gate	*puc*	$0.5	$2
Sequential depth	*seqdepth*	0	103
Production volume	*vol*	1000	1 000 000

Figure 3.4 shows the sensitivity for each parameter. The sensitivity factor was set to 1.1, which refers to a 10% variation of the parameter. The vertical line marks the range in which the sensitivity will be in 99% of all cases. The small horizontal bar marks the mean value of the sensitivity. The figure presents the results for the test method *no DFT*. Figures 3.5 and 3.6 provide results for scan path and circular self-test path respectively.

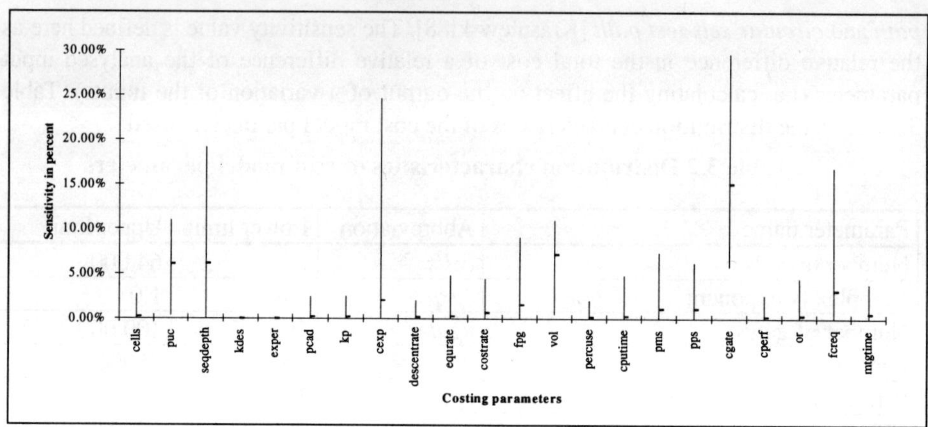

Figure 3.4 Sensitivity of total cost with a sensitivity factor of 1.1 (10% variation), no DFT

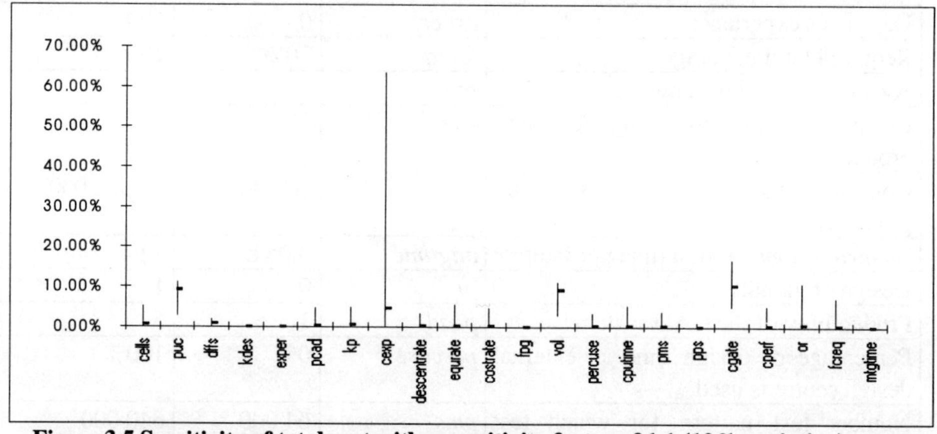

Figure 3.5 Sensitivity of total cost with a sensitivity factor of 1.1 (10% variation), scan path

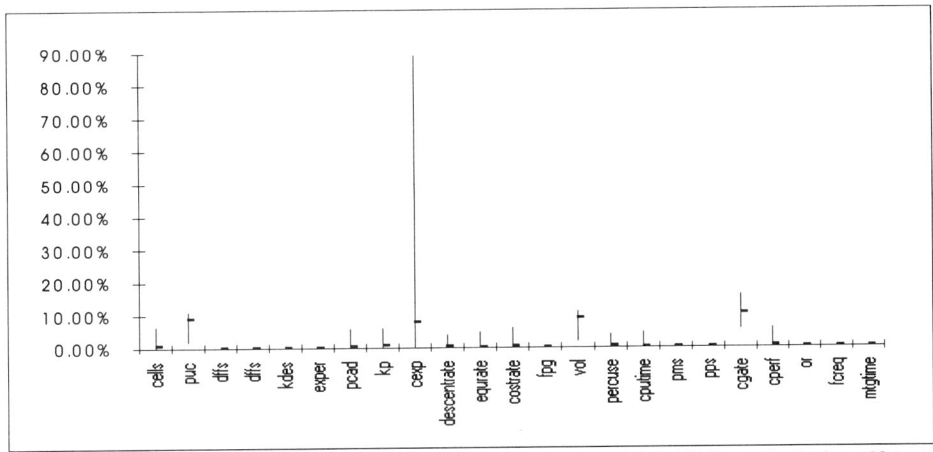

Figure 3.6 Sensitivity of total cost with a sensitivity factor of 1.1 (10% variation), self-test

We distinguish between the three cases no DFT, scan path and self-test. The analysis was performed using 10 000 simulations. For each case the parameters are classified as shown in table 3.3.

Table 3.3 Parameter classification according to sensitivity

Class	No DFT	Scan path	Self-test
High mean, high maximum	*cgate, seqdepth*	*cgate*	*cexp,*
High mean, av. maximum	*puc, vol*	*puc, vol*	*cgate, vol, puc*
Av. mean, high maximum	*cexp, fcreq, fpg, costrate, pms, pps*	*cexp*	
Low mean, av. maximum	*cells, pcad, kp, descentrate, equrate, percuse, cputime, or, cperf, mtgtime*	*cells, or, fcreq, pms, cperf, cputime, pcad, kp, equrate, costrate, descentrate, percuse,*	*cells, pcad, kp, descentrate, equrate, costrate, percuse, cputime, cperf*
Low mean, low maximum	*kdes, exper*	*dffs, kdes, exper, fpg, pps, mtgtime*	*dffs, kdes, exper, fpg, pms, pps, or, mtgtime*

From this classification and figures 3.3 to 3.6 the following conclusions can be drawn:

- The parameters *number of gates*, *sequential depth*, *production unit cost per gate* and *production volume* are the most important parameters for the sensitivity and accuracy of the cost estimation.
- The parameters *constant factor concerning designer's productivity*, *designer's experience*, *number of cells*, *productivity of the CAD system*, *constant factor concerning productivity*, *design centre cost rate*, *computer equipment rate*, *percentage of design time an external design centre is used*, *CPU time*, *design originality*, *performance complexity* and *manual test generation time per fault* do not affect the sensitivity of the total cost very much in most cases. Inaccurate data for these values may fulfil the accuracy requirements of the total cost.

The sensitivity to the total cost for some parameters remains negligible in at least 99% of all cases. This leads to the question, whether the total cost is sensitive to these parameters at all. To answer this question, the maximum sensitivity for the parameters must be determined. If the maximum sensitivity is negligible, the cost model can be refined by taking out the insensitive parameters. The determination of the maximum sensitivity is an optimisation problem. As we are using the Monte Carlo method, we had to implement a method to reduce the sample size of the Monte Carlo simulation, because crude Monte Carlo simulation [Hammersley65] would lead to an extremely large sample size, and implemented a technique called *importance sampling.* This enabled the calculation of the maximum sensitivity conditions with only 1000 simulations.

The maximum sensitivity was determined for the following parameters: *performance complexity, design centre cost rate, designer's experience, average number of faults per gate, constant factor concerning designer's productivity, constant factor concerning total productivity, manual test generation time per fault, number of flip-flops, productivity of the CAD system, percentage of design time an external design centre is used.*

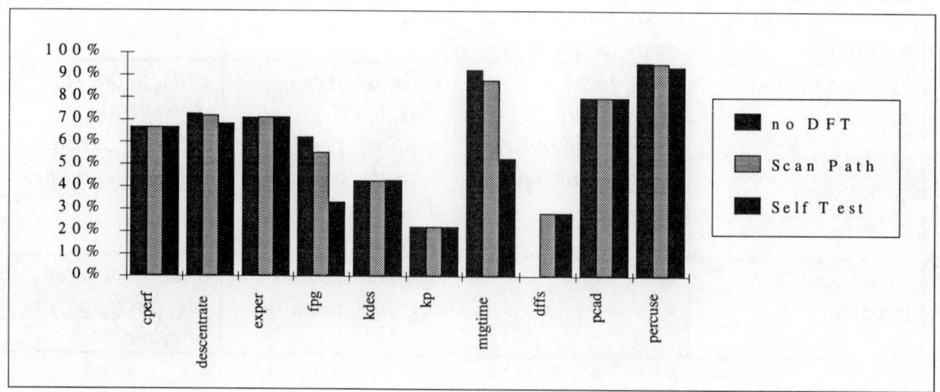

Figure 3.7 Maximum sensitivity values

The analysis was made for the test strategies *no DFT*, *scan path* and *self-test*. The resulting maximum sensitivity as a percentage of the related maximum cost is shown in figure 3.7. This shows that the maximum sensitivity exceeds 20% for all parameters, which is by far too high to neglect. This means that for all parameters situations may occur where a variation of the parameter within the range defined in table 3.2 can lead to more than a 20% variation of the total cost. A simplification of the cost model in general by removing parameters is therefore not possible.

3.5.3 Iterative sensitivity analysis

This second application of sensitivity analysis leads to a test strategy planning procedure in iterative steps as follows:

1. Make a rough and inexpensive estimate of the input data.
2. Perform test strategy planning.
3. Perform a sensitivity analysis in order to find out which parameters need to be defined more accurately. These are those parameters which have a high sensitivity.
4. If the accuracy of the result is OK, then the test strategy planning procedure is finished, else make more accurate estimates for the most sensitive parameters and go to step 2.

In the iterative sensitivity analysis, the same methods are used as for the general sensitivity analysis. The main differences are the iterative procedure and the different distribution of the parameter values. For the general sensitivity analysis, we used a uniform distribution for all parameters in the range of all possible values. For the iterative sensitivity analysis, we have inaccurate estimates for the parameters, which can be either uniformly or normally distributed.

3.5.4 Total variation sensitivity analysis

This third sensitivity analysis application considered the trade-off between data acquisition effort and the accuracy of the result. In practical terms, there may be a certain cost limit for data acquisition, which makes estimated input data remain inaccurate. This means that the decision criterion, i.e. the resulting cost value, is subject to inaccuracy. This inaccuracy will vary with the test strategy. The following example will illustrate this fact.

Comparing the test method 'scan path' in combination with combinational ATPG to 'no scan path' in combination with sequential ATPG, the related test generation costs as part of the resulting cost are much more uncertain for 'no scan path'. The reason for that is the inaccurate parameter 'sequential complexity', which affects sequential ATPG cost but not combinational ATPG cost. This may lead to a situation where the resulting costs for the sequential ATPG approach are lower than those for the combinational ATPG assuming that the estimated value for the sequential complexity is exact. At the same time, there is a high probability that the costs will be much higher for the sequential ATPG approach, considering the high probability that the sequential complexity will be higher than estimated. In order to take the inaccuracy of the cost estimation into account, the total cost value was replaced by the distribution function of the total cost value as the optimisation criterion. This provided a mechanism for a trade-off analysis.

In summary, the general sensitivity analysis showed that for each parameter configurations may occur which lead to a maximum sensitivity of more than 20% of the total cost. But in most cases, the total cost is sensitive to only a few parameters of the cost model. These are the number of gates, the sequential depth, the production unit cost per gate, the production volume and the complexity exponent. Based on this outcome, it is proposed that economics based test strategy planning is performed as follows:

- Provide values for the 'important parameters' listed above.
- Perform iterative sensitivity analysis until the accuracy requirements are fulfilled.

This method can drastically reduce the effort spent in data gathering, as in most cases the accuracy requirements can be fulfilled by providing well defined values for only a few parameters.

4

The ECOtest system and automatic test strategy planning for ASICs

4.1 INTRODUCTION

The EVEREST test strategy planner (ECOtest, for economics based testing) was developed under the 'Test Economics Modelling' task of the ESPRIT EVEREST project. This was a collaborative project between several major European companies and universities, with the aim of producing an integrated set of tools for testing VLSI chips as well as VLSI based systems. The tools included a verification tester, a test specification format, a set of test generation and validation tools, design for testability for a silicon compiler, and knowledge based solutions to test strategy planning. It was the last objective that led to the development of ECOtest.

The aim was to create a test economics model and use it in a test strategy planning system. A high level test strategy planner was an essential part of the design for test philosophy of the EVEREST project as a whole, especially as the step prior to the integration of design for testability methods into a silicon compiler. The innovative, as well as the collaborative nature of the project, created a set of requirements for the final system:

- the test strategy planner was intended for use by industry, and therefore had to take into account industrial practices and needs
- it should provide a novel approach to the solution of the test strategy planning problem.
- the test strategy planner needed to fit into a suite of tools, and be evaluated and used by several different companies. As a result, the system had to be flexible and adaptable to different users' needs without major reprogramming.

Brunel University was a partner to Siemens-Nixdorf Informationssysteme in the design and implementation of the project. In this type of project, there are several advantages to be gained from academic-industrial collaboration. The lack of reliable industrial data can become a serious drawback to systems developed in the academic sector. Companies can be reluctant to part with what may be considered sensitive information, or cannot allocate the time and resources to find the data needed for successful economics modelling. Close academic-industrial collaboration largely removes these problems. It allows for an accurate specification of the problem, based on the requirements of the actual users, and data are available, as are the resources to find out any information that is not readily to hand. Cost model equations can be validated from previous experience, and models that closely reflect the practices of the company can be created. From an academic point of view, industrial and human resources for the implementation of ideas can be an important advantage. In a test strategy planning system, a large amount of preliminary implementation work has to be done before ideas can be tested and different test strategy planning methods evaluated.

In the rest of this chapter, the philosophy of ECOtest will be discussed first. The architecture of the system will then be described, and the structure of the constituent parts examined in detail. In particular, a number of methods for automatic test strategy planning will be discussed with the aid of real circuit examples.

4.2 THE ECOtest PHILOSOPHY

The ECOtest test strategy planning philosophy is based on the fact that there are important advantages in early consideration of design for testability issues, and that a structured, automated system is needed to advise the designer as to the options available in terms of design for test methods, and provide the means for an evaluation of the options. An evaluation in economic terms was considered to be the only way that different design for test methods could be compared objectively. The objectivity of test method evaluation is a serious drawback to other test strategy planning systems such as TDES [Abadir85], where a scoring system is used, based on the user's understanding of the relative importance of test parameters.

The test strategy planner was developed specifically for the design of ASICs, designed using a CAD system with predefined library cells. It was assumed that chips were fabricated by an external vendor, who would also test them with test patterns supplied by the designer. This assumption affected only the cost modelling, and a different, more complex cost model can be written to handle in-house fabrication. Test pattern generation would be the designer's responsibility. The test strategy planner relied on the process of block structured design using predefined library cells to obtain partitioning information. Optimised automatic partitioning for test strategy planning is a complex problem, which was examined but not attempted in this instance. The system required, however, that testable units be homogeneous, i.e. consist of only one design type.

The definition of the application area of the test strategy planner affected the design of the cost model. For example, the fact that an external vendor fabricates the chips means that the cost of production can be easily calculated from the vendor's prices, if a good estimate of equivalent gate count can be produced. Yield effects need not be calculated explicitly, as they have already been taken into account by the vendor in pricing the ICs.

The cell based design method had the advantages that information on predefined cells could be acquired directly from the CAD tools through links with the design libraries. For example, if the logic overhead of a test method is given in terms of D-type flip-flops (DFFs), the actual gate overhead can be calculated once the equivalent gate count of a DFF is retrieved from the design library. As the system had to be flexible enough to cope with different levels in the design process, especially considering that different blocks of the design may be at different stages, the design description had to be fully hierarchical for every block. It was deemed that the cost model was sufficient for any level.

ECOtest used stored test method information for the formulation of test strategies. In order to provide accurate estimates of the economic effects of a test method (such as gate overhead and test pattern number) the stored knowledge could be in the form of equations, evaluated during the test strategy planning process, using data supplied by the CAD system. The whole system had to be portable, maintainable and usable by a number of different companies. Its ability to integrate into a larger suite of programs was also an important consideration. Speed was also significant, as the prototype was to undergo industrial evaluation. The software was implemented in C.

The ECOtest system intended to provide the designer with a set of tools to allow the evaluation of test methods and generation of test strategies. However, this can be a long process if done manually, especially with large chips. For that reason, a set of algorithms were developed to automate the process. As an exhaustive evaluation of all test strategy combinations is not a practical undertaking for most circuits, the algorithms attempted to reduce the search space by heuristic means.

ECOtest has a CAD tool link, and therefore has the potential to implement a chosen test strategy. However, it was decided that the test strategy planning system would act as an adviser to the designer, and not force design changes or create a test program plan. This route was chosen because in the early stages of the design there is often not enough information to allow reliable test method embedding in the design description, and also because it is intended that the system can be used at different stages, gradually refining the test strategy as more information becomes available. It should be kept in mind that the economics model and the test strategy planner work on *estimates* of the effects of test methods, *before* these methods have been implemented. It is up to the designer to implement a test strategy, using information provided by the test planner on implementation guidelines. The final optimisation can take place at that stage.

4.3 THE ECOtest ARCHITECTURE

Figure 4.1 shows the outline of the test strategy planning system. There are two distinct parts of the system, which operate independently but have access to a common set of files. These are the design specification reader and the test strategy planner. They are kept separate as their actions are distinct and do not overlap in any way.

The design description was acquired either directly from the user or from an existing netlist, and was built in a hierarchical fashion in order to allow test strategy decisions to be made at several stages of the design process. The design specification reader also had access to CAD library data, for calculating the sizes of functional blocks. Other essential economic data which could not be directly acquired from a netlist description were requested from the user. At the end of this process, a design description file was created, which was stored and used to initialise the cost model for the design, ready for the test strategy planning. The design specification reader also allowed the user to alter an existing design.

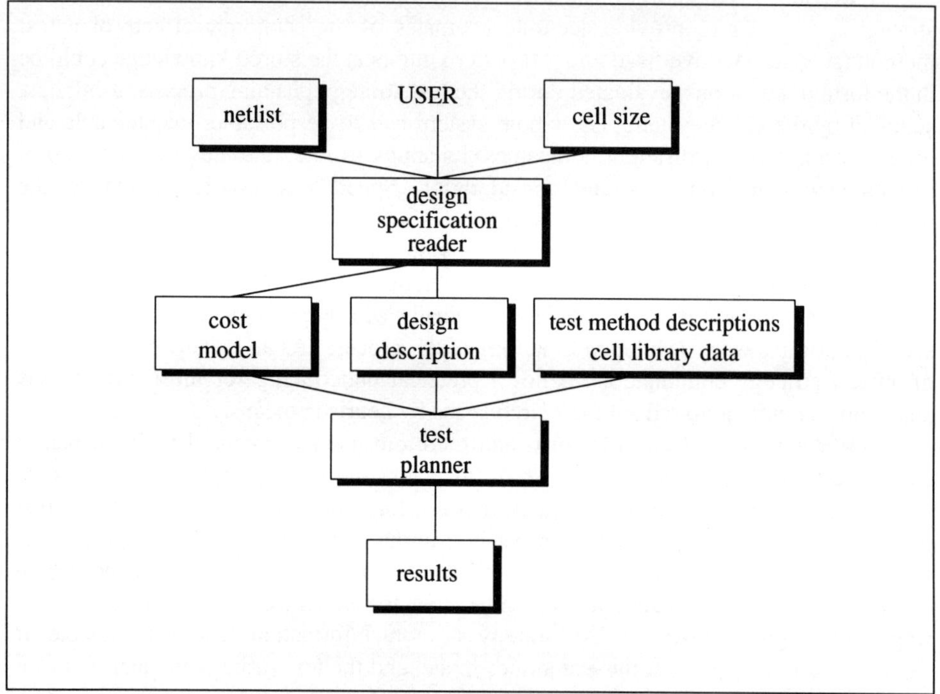

Figure 4.1 The ECOtest architecture

The data were used to update an economics model automatically prior to the test strategy selection process. The design specification reader would normally only be run once, and a design description file created. That file would then be accessed by the test strategy planner.

The test strategy planner contained the programs for the evaluation of the cost model, as well as the evaluation of the effects of test methods on the design. It took its input from the design description and cost model files, and used stored knowledge on test methods to evaluate a variety of test strategies. Cell library data were linked to test method descriptions. The test strategy control could be left entirely to the user, and a group of functions were provided to apply test methods and examine the cost model and the testability status of the circuit. Alternatively, the user had the option of employing one of a set of algorithms that performed automatic test strategy planning to accelerate the selection process. Both manual and automatic test strategy planning runs could be stored and restarted at any point, and the user also had the option of starting the selection process manually, and continue by using one of the automatic planning algorithms, or vice versa. Results could be stored for later reference. Data and programs were kept separate for greater flexibility in maintenance and also because in this way different users could make modifications or additions to the cost model and the test method descriptions without reprogramming.

4.4 DESCRIPTION OF TEST METHODS

The test method description data base had to take account mainly of the economic effect of test methods, so that an economic evaluation could be produced, but also of the suitability of test methods for the particular design so that fundamental requirements (in terms of fault cover, gate count, maximum pin count, etc.) would still be met. Therefore the data needed to describe test methods were mostly dependent on the cost model parameters that would be affected by a design for test method.

Parameters in the test method data base fields were described either as single values or text strings, or as equations. The parameters described were categorised in three groups, and are shown in table 4.1.

Table 4.1 Test method data base description parameters

Group 1: Parameters which are used by the cost model for an economic evaluation.	
Equivalent gate count	Number of extra functions
Sequential depth	Number of test patterns
Performance complexity	Achievable fault cover
Additional pin count	TPG method
Originality	Pin compatibility (possible shared use of test pins)
Group 2: Design implications.	
Accessibility impact	Self-test / no self-test
Test method type (testability/accessibility)	
Group 3: Design requirements.	
Suitable design class (e.g. PLA, RAM)	Suitable design style (e.g. synchronous, flip-flop design, latch design)

The first group of parameters were used in the cost model for an economic evaluation of the method. The performance complexity and the originality parameters were related to design costs, as was the number of extra functions (if any) introduced by the test structures. The pin compatibility information was used in the calculation of the final pin overhead of the plan. It indicated whether the extra pins needed for the test method could be shared. An example is the scan-in input for a testable unit, which could be used for the scan path of another testable unit if the scan paths were in the same chain or multiplexed. Also the test pattern generation method was indicated, which could be manual, automatic using a combinatorial ATPG, automatic using a sequential ATPG, pseudo-random, or exhaustive. The effective average sequential depth was also indicated, as it could be affected by the method (scan path makes sequential logic appear combinational).

A set of equations described the minimum achievable fault cover, the logic overhead in terms of equivalent gates, the extra input and output pins needed on the chip, and the number of test patterns. The user could specify the design variables in the equations, such as the gate count of the block or the number of D-type flip-flops, which could be retrieved from the design description or the cell library data base. In this way, parameter values could be calculated for different implementations. In order to increase the flexibility of the system, the user could also define new identifiers, which the system would request at the time of test method application. For example, with some pseudo-random test methods, the number of times the self-test sequence is repeated was variable according to the fault cover required [Illman86]. By default, the system calculated the number of repetitions needed to achieve the user defined minimum fault cover requirement, but a user could evaluate different possibilities by setting the repetitions value to be a user defined identifier.

The second group of parameters described the design implications of the test method. The test method type was defined in this group, and the accessibility effects of the method were taken into account. A requirement of the system was that, at the end of the test strategy planning process, a test strategy should make every block testable in isolation, and every block i/o line accessible, so that test patterns could be propagated to it. The test methods were therefore categorised into testability enhancing (internal) and accessibility enhancing (external), in order to classify the different effects they have on the circuit.

A testability enhancing method (also called an internal method in the test method description) improves the testability of the block it is applied to. This normally means a guaranteed fault cover. A side effect of this may be that the structures used can provide access to the i/o ports of the block as well. An example of such a method is a scan path which improves the testability of a highly sequential block. If the scan path also includes the i/o pins of the block (through i/o registers), it can be accessed to and from the primary i/o of the chip. This would be classed as a testability enhancing method, as would a self-test method that does not necessarily facilitate access to the block from the primary i/o.

An accessibility enhancing method (also called an external method in the test method description) has the sole purpose of improving the accessibility of the block, without necessarily affecting its testability in any way other than by virtue of the fact that test patterns can then be propagated to the block to test it. A scan path which includes only the i/o lines of a block would fall into this category. The definitions, therefore, partly overlap, in that an internal method can also imply accessibility. External methods are used to resolve any accessibility problems remaining at the end of the test strategy planning process. Accessibility flags can be propagated through design blocks via transparent paths. These would be defined by the user when the design description was entered, and contained information on paths *known* to be transparent; for example, the path from the input to the output lines of a RAM can be considered transparent.

It was obviously not adequate simply to classify a method as testability or accessibility enhancing. It was essential also to specify exactly its accessibility effects, if any, on the block's i/o. As structured DFT methods were used, it was enough to specify the type of i/o line the test method made accessible. Therefore, five fields were defined to hold accessibility flags: accessibility for data inputs, accessibility for control inputs, accessibility for clock lines, accessibility for outputs, and accessibility for bus (bi-directional) lines. For the purposes of the test strategy planner, these categories were adequate to categorise the accessibility effects of both internal and external test methods.

The third group of parameters contained the design requirements of the test method. For a test method to be applicable to a block, it had to be suited to the type and style of the design and also needed to fulfil basic requirements in terms of fault cover, maximum gate count and pin count. Data in this section were used to check that design requirements were met and contained fields of the design style and design type the method was applicable to. Note that a single method may be applicable to multiple design types. The design type and design style categories are shown in table 4.2.

Table 4.2 Design types and design styles currently supported

Design types	Design Styles
Sequential random logic	Synchronous latch design
Combinational random logic	Synchronous flip-flop design
PLA	Asynchronous design
RAM	
ROM	
Multiplier	
ALU	

4.4.1 Test method description - an example

The rest of this section provides an example of the coding of the test method description. The method described here is a self-test PLA design [Fujiwara81]. The method employs the idea of parity testing. An extra product line plus its crosspoint connections can be added to ensure an even or odd number of bit lines. A single-crosspoint fault in the AND array would change the parity of the bit lines. Faults in the OR array can be detected by adding an extra output line. This method employs two parity checkers and a universal test set, requiring storage for test vectors, or the ability to generate them on chip. Although the PLA itself is self-testable, the method does not necessarily require or provide accessibility to and from the primary i/o. Table 4.3 shows the test method description in the data base.

Table 4.3 A test method description example

Test method name	Fujiwara
Test method type (ext./int.)	internal
Suitable design classes	PLA
Self-test	yes
Assures data-in accessibility	no
Assures control-in accessibility	no
Assures clock-in accessibility	no
Assures data-out accessibility	no
Assures bus accessibility	no
Performance implication	1.1
Test pattern generation method	none (universal test set)
Sequential depth	0
Achievable fault coverage	1
Number of test patterns	(3+2log$_2$ * number_of_product_lines)*(2* number_of_inputs + 3 * number_of_product_lines)
Overhead formula	(4*number_of_inputs + 11* number_of_product_terms + 5*number_of_outputs + 6)/4
Originality impact	0
Inpin overhead	3
Outpin overhead	2
Bipin overhead	0
Pin compatibility class	class 32
Design style requirements	asynchronous, synchronous latch or synchronous flip-flop
Number of functions	0

Figure 4.2 shows the cost model operation during the evaluation of the effects of a test method on a block. A set of parameters such as gate and pin count and number of test patterns are precalculated before being merged into the model with the other test method data. The model is then recalculated. Once the test method data are in the model, the values for that block are not altered until another test method is applied to the same block. This ensured that any effects (such as the test vector number) that the application of a given method for a block might have on the costs for other blocks were taken into account.

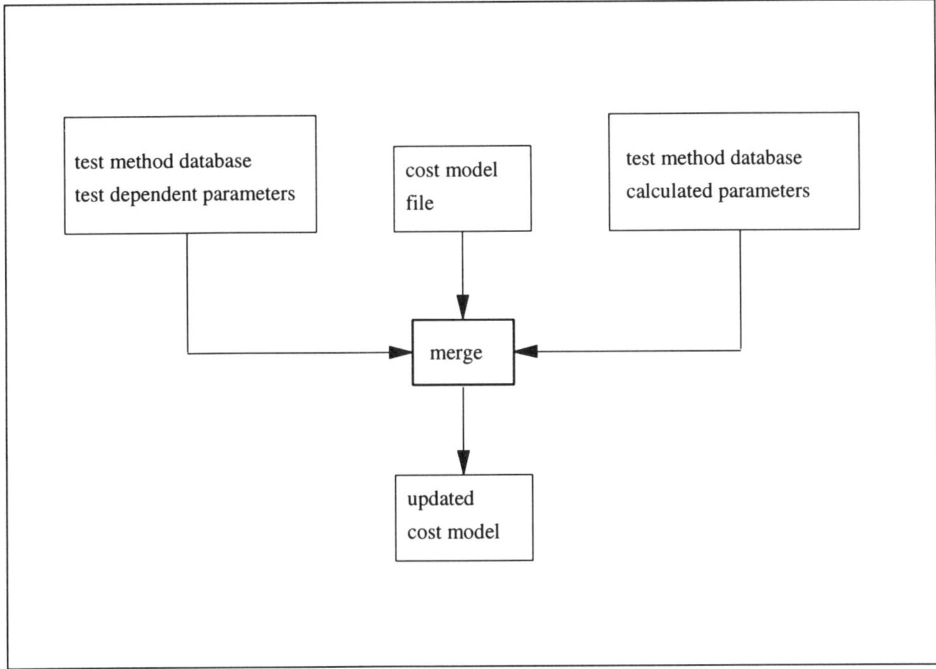

Figure 4.2 Updating the cost model

4.5 THE TEST STRATEGY PLANNER

This section describes the requirements and implementation of the manual test strategy planning part of the ECOtest system, as well as the user interface functions of the test planner. The effects of a test method on the accessibility status of individual blocks as well as the circuit as a whole are discussed in detail.

4.5.1 Test strategy planning requirements

The minimum requirement for the test strategy planner was that it should provide the user with functions that allowed the effects of the application of a test method to a given block to be calculated and examined, in terms of cost and design implications. As a test method may affect more than simply the block it is applied to, global effects should be considered as much as possible. The process should be iterative; that is, it should be possible to apply a selection of test methods to a block independently of each other. In essence it should be possible to 'undo' the effects of a test method when another is applied to the same block.

The test planner then should be able to check the applicability of a given method to a specific block. This is necessary, as some methods are either specifically designed or at least optimised for specific blocks, and are not necessarily applicable to others. An example is a range of test methods specific to PLAs [Zhu88]. Other methods, such as scan path, require a synchronous design style. In order to calculate both cost and design effects, the test planner must have access to the stored design description, initial cost model and test method descriptions, and have facilities for cost model updating and evaluation. The data in the test method description must be used to update the cost model of the whole design, so that global economic effects are taken into account.

As testability and accessibility methods are considered separately, the test strategy planner should also have the ability to calculate the accessibility effects of a test method, not only on the block it is applied to, but on the whole design. It should also be possible to apply external (accessibility enhancing) methods in the most economical way for solving accessibility problems. Again the application of test methods must be reversible. As the test strategy planner acts in an advisory function, any changes to the design description should be temporary, current only for the duration of the test strategy planning process. This will allow a variety of test strategies to be considered, based on the same design description. As the planner will ideally be used in the early stages of the design, it was not considered a priority to implement the test methods in the design description, as in most cases the necessary design data for detailed implementation would not be present. It is then up to the designer to implement the test structures suggested, where a degree of further optimisation is likely to happen. Some further information on the test methods described should be available to the designer in text form to help this process. One of the intended uses of the test strategy planner is to provide an initial test strategy for a silicon compiler, where the actual implementation would take place.

4.5.2 The test strategy planning process

This sub-section describes the actions taken by the test planner in order to estimate the effects that a test method applied to a specific block has on the design as a whole. These actions are the same whether manual or automatic test strategy planning is used.

The first requirement was that the test method is compatible with the testable unit under consideration. Therefore, before any evaluation was done on the cost model, the following conditions would be checked:

1. Is the test method valid? A description of the test method must exist in the data base.
2. Is the block under consideration a testable unit? The system provides the option of considering blocks not to be testable units. This can be useful if, for example, the block controls a critical timing path or the designer would prefer to use an *ad hoc* design for test strategy for it. In that case, the block is excluded from test method application.

3. The design style of the testable unit must correspond with the design style field of the test method considered. Test methods are specific to testable unit types, in order to take advantage of specially optimised methods and fault models, and maximise fault coverage for different types of fault.
4. Similarly, the design type of the test method must correspond to that of the testable unit, in order to establish whether the test method is applicable to the particular implementation of the block.

In addition to these basic requirements, the user could set certain limitations on the finalised test strategy. These took the form of a maximum equivalent gate count, a maximum number of pins, and an upper bound on self-test time. If any of these limits were exceeded, the test method application failed and the test method was reset. However, these requirements could not be checked prior to cost model evaluation. Therefore, there was the possibility that a test method application would fail, even though the initial conditions prior to cost model evaluation were satisfied. A minimum fault cover achieved was another requirement set by the user, but that value was used only to calculate the cost of achieving the specified cover.

The test method application process would proceed as follows. Once the basic requirements are met, the test method description data that are used to update the cost model are retrieved, and the cost model for the design updated and evaluated. The updating could take several forms; for example, a calculated or constant overhead might be added to a cost model value, such as the logic or pin overhead. A value in the cost model can be replaced with one from the test method description, as would be the case in a method that guaranteed a certain level of fault coverage. Note that the initial cost model would be retrieved from the design description. That would be initialised with a default set of test methods, as a baseline for further calculations. The user could set a default test method for each type of testable unit. A predefined test method is essential, as the initial cost model needs to be valid in all respects. A default method for a RAM, for example, would specify the algorithm used in deriving the test patterns and therefore the test pattern number. Default test methods were coded in the standard form in the test method descriptions data base.

So far the accessibility effects of the test method have not been mentioned. These are necessary in order to determine whether a test strategy fulfils the system requirements of both testability and accessibility. Therefore, after a test method is evaluated, the accessibility status of the circuit is recalculated. This was done globally, as the accessibility status of a set of lines may be propagated through transparent paths to other parts of the circuit. Accessibility calculations take the form of propagating accessibility 'tokens' from the primary inputs to the primary outputs of the circuit and vice versa, taking advantage of transparent paths where they exist. If a line bundle was made accessible as a result of a test method application, it was marked as primary i/o, and therefore both observable and controllable.

Accessibility tokens consisted of a controllability (access from the primary inputs) and an observability (acccss to the primary outputs) flag. Both should be set if a line was to be considered accessible. This controllability and observability definition is different to that of testability analysis systems, in that it only assumes a true/false value.

Intermediate values were not considered, as the objective was to guarantee accessibility rather than estimate the probability of achieving it. A full testability analysis was not practical for the level of detail likely to be present in the planner, but the use of accessibility flags allowed the user to gain an understanding of the test method implications for the testability/accessibility of the whole design.

When a test method was applied, its effects on the accessibility of the block were also noted, and the accessibility of other block i/o recalculated. The recalculation was based on existing information of transparent 'paths' through functional blocks, and used a recursive depth-first algorithm to propagate accessibility tokens through the circuit. Thus, and by the use of the cost model, global as well as local implications of the test method were assessed. Testability and accessibility enhancing methods were applied separately, and each block could have a maximum of one testability and one accessibility method allocated to it at any one time. The system provided the necessary functions for the user to evaluate a variety of test strategies, and also to use a degree of automated planning. Automatic test strategy planning is discussed in detail later in this chapter.

The cost implications could be examined by examining the cost model. At the end of the process, a test strategy description was produced, with details of the test method chosen for every block, together with a complete copy of the cost model for later reference. The design description itself was not altered.

There were some differences in the application of testability and accessibility enhancing methods. Testability enhancing method application followed the procedure described above. However, accessibility enhancing methods were applied to resolve specific accessibility requirements. As a result, a scan path, for example, might not be necessary for all i/o lines of the block in question, but only for those with accessibility problems. An accessibility analysis would first be carried out, and an external method applied only to those lines with accessibility problems. This method was used in order to avoid the application of redundant test structures, and take into account existing test structures which would affect the application of subsequent methods.

Therefore, although a test method primarily would affect the block it was applied to, global effects were also taken into account through the cost model and accessibility analysis. Although the design was treated as a collection of distinct blocks, the use of a global economics model for the whole of the circuit ensured that some measure of global knowledge was used. This was different to the way that, for example, TDES [Abadir85] compares designs, where the comparison is on local (block) level, and may not necessarily provide a realistic comparison. It has been shown, for example, that the overall size of the chip in which a testable unit is embedded can affect test method trends, and possibly the selection of the optimal test strategy [Dislis89]. In ECOtest, optimisation considerations dynamically affected the formulation of the test strategy.

4.5.3 The test strategy planner functions

This sub-section briefly describes the commands available to the user with which test strategies for the design can be formed and evaluated, in order to illustrate the path a manual test strategy planning process can follow. The complete set of test strategy planning commands available in ECOtest is shown in table 4.4. The 'apply' command allows the user to apply individual test methods to a block. One testability and one accessibility method may be applied to a block at any time. This command invokes the test method application process described in detail in the previous section. A set of automatic test strategy planning commands invoke a number of different test strategy planning algorithms.

Table 4.4 ECOtest commands

Command	Description
apply	apply test method to a given block
auto_plan 1	automatic test strategy planning (basic algorithm)
auto_plan 2	order blocks by accessibility before test planning
int_methods	apply cheapest internal method to block
ext_methods	apply best combination of external methods to circuit
optimised_ext_methods	re-evaluate circuit after external method application for further optimisation
sim_anneal	simulated annealing for test strategy planning
exhaustive	exhaustive test strategy evaluation (only for small circuits)
access	show line accessibility
show_tp/show_ci	show test strategy/circuit information
stack	place cost model values on stack
report	print cost data of the test strategies on the stack
dump_cm	print evaluated cost model
file 'filename'	store cost model to specified file
new 'parameter' = value	evaluate the cost model using the new value for the specified parameter
tell 'parameter'	print the value of the specified parameter for test strategies on the stack
expand 'parameter'	expand a parameter to its primary parameters
table	create a table of the values of one parameter based on variations of another (used for graphical output).
q	quit

Accessibility analysis is performed as part of the test method application process, and the accessibility status of lines can be seen using the 'access' command. This function displays all testable unit connections (bundles), which are not controllable or observable by either primary i/o's, the transparency of neighbouring testable units or an applied test method. The function is based upon the design data and the pre-calculation of the accessibility implications. This command is provided to inform the user about parts of the design where the accessibility is limited, in order to target these parts for further investigation. Accessibility is given as a yes/no flag, and may take the form of accessibility from the circuit's primary inputs ('controllability'), or accessibility to the circuit's primary outputs ('observability').

The main actions of the manual test strategy planning commands are shown in figure 4.3.

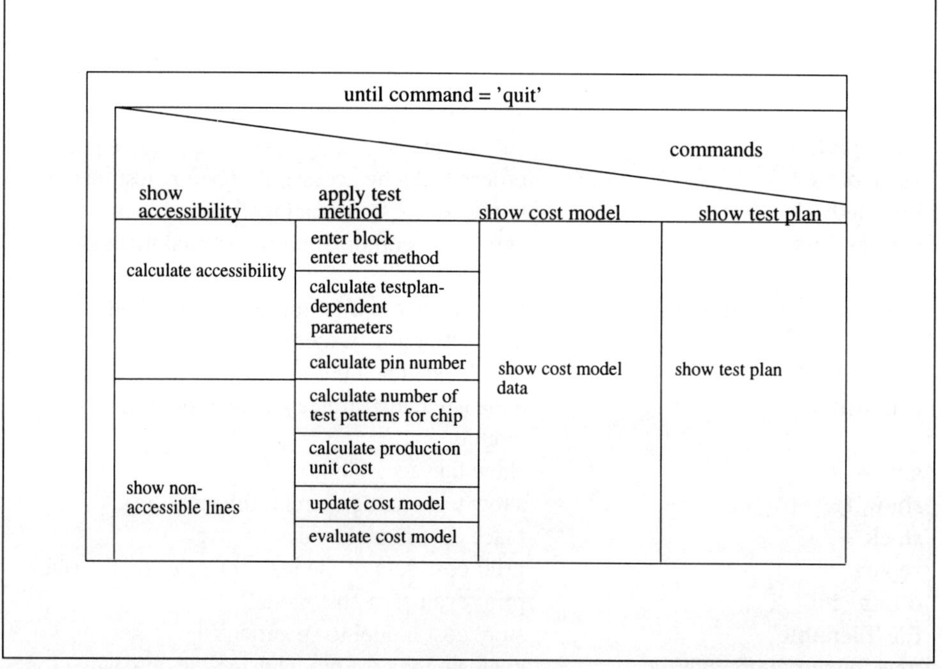

Figure 4.3 Manual test strategy planning

The rest of the commands allow the user to examine the test strategy in detail, in order to provide guidance towards an optimal strategy. The functions available are either information functions, or cost model manipulation functions. The user can request circuit information using the 'show_ci' (show circuit information) command, as well as information about the chosen methods for each block using the 'show_tp' (show test plan) command.

The cost model data can be shown on the screen using the 'dump_cm' (dump cost model) command, or stored in a file using the 'file' command and a specified filename. It is possible to compare up to five strategies and their cost model data by successively placing each one on a stack ('stack'). The cost model data for all strategies on the stack can be seen by using the 'report' command. Parameter values can also be examined individually by invoking 'tell'<parameter_name>. It is also possible to assign new values or equations to individual parameters ('newequ' <parameter_name> = <value>), or alter the value of a parameter through a specified range and observe the result on another parameter graphically using the 'table' command. An example of this would be when the user would like to evaluate the effect that production volume would have on the relative costs of design, manufacture and test. It is also possible to find out which factors a parameter is dependent on, by using the 'expand' function, which provides the equation of a secondary parameter in terms of primary values.

Summarising, the user was provided with tools to allow evaluation of test methods in terms of cost, to estimate the effects on the design itself, and to examine and manipulate the cost model in order to keep track on the development of a test strategy for the circuit as a whole. Many of the functions described above were also useful in the development of automatic test strategy planning algorithms.

4.6 AUTOMATIC TEST STRATEGY PLANNING FOR ASICs

In ECOtest, the problem of automatic test strategy planning was one of allocating a set of predefined test methods, with predictable economic and design effects, to the functional blocks in the design. The test method allocation had to be economically optimal, that is, the overall cost as calculated by the economics model had to be minimised. In addition, a set of user defined constraints had to be satisfied. These constraints included limitations on the minimum fault cover to be achieved, maximum gate count and number of pins, and maximum self-test time.

The choice of test method for a specific block may influence subsequent choices for other blocks in the design. For example, the introduction of scan path at the i/o of a block may remove the need for scan path at the i/o of an adjacent block. Additionally, any register which can be configured to act as a scan path (such as a BILBO), may also fulfil this requirement. The exact configuration would depend on the blocks' connectivity. The problem is one of combinatorial optimisation, and the total number of possible solutions increases factorially with circuit size.

An exhaustive search for the best solution was only practical for very small applications, and rapidly became impractical for realistically sized designs. It was therefore essential to find alternatives to an exhaustive search. The alternatives described in the rest of this chapter exploit the decision making structure of the economics model as well as the circuit design representation to create an optimal test strategy. In creating the algorithms, the scope was narrowed from the general combinatorial optimisation problem to the much more specific problem of test strategy planning for ECOtest. By sacrificing some of the generality of the solution, it was possible to create answers to the test strategy planning problem which took practical considerations into account and were tailored to the system.

In the following sections, the requirements of test strategy planning algorithms for ECOtest will be described, with reference to previous work in the field. A range of algorithms providing solutions to the problem will then be discussed, with the aid of an example circuit. Finally, the use of simulated annealing for test strategy planning will be addressed. This was an innovative approach, not previously used for test strategy applications.

4.6.1 Requirements of the test strategy planning algorithm

Previously published work on the subject of test strategy planning was reviewed in chapter 2. It was noted that the system closest in its objectives to the ECOtest test planner, although by no means identical, was the TDES/TIGER system [Abadir85, Abadir89]. It adopted a broadly similar approach to partitioning the circuit into functional units called kernels and evaluating the effects of test methods, termed testable design methodologies (TDMs). The test strategy planning actions of TDES/TIGER are those of TDM embedding and test plan generation. TDM embedding matches TDMs to kernels (the partitions in the design). The TDM embeddings are evaluated according to a scoring function, which includes test time and area overhead effects, as well as fault cover. These are weighted according to their relative importance to the user, but also have to satisfy user defined constraints. Backtracking is allowed to resolve resource sharing conflicts, and also is a result of dynamic adjustment of control weights by the user to satisfy the constraints.

TDM embeddings and their cost functions are evaluated separately for each kernel. The only allowance for global optimisation is that built-in test structures with a high 'shareability' rating are preferred. These are structures that can potentially be used by a number of different kernels, as they appear in several kernel subgraphs. However, there is no guarantee that this shareability *will* be used; it is only used as a heuristic in the TDM embedding process.

There are several drawbacks to using a similar approach in the ECOtest system. The most obvious drawback is the subjectivity of the scoring function, which hinges on user specified relative weights for important parameters. This approach relies heavily on the user's experience and knowledge of factors which may fall outside his/her direct responsibilities, and can lead to extensive backtracking, as the system will allow a user to enter different scoring functions in order to allow certain solutions to be evaluated. The test strategy planning process therefore is a combination of the system trying to satisfy user constraints and suggesting changes to these constraints when solutions prove difficult to find. There would be no need for constant constraint re-definition if the objectivity of the scoring function was accepted.

In the ECOtest test strategy planner, the use of a detailed economics model solved this problem in that, once the initial investment of creating the economics model for the user company is made, an objective measure of the relative importance of parameters in terms of cost is readily available. Once the cost model is defined, the scoring function stays the same for the duration of the process. Therefore there is no need to backtrack because of changes in the scoring function.

Although it may not be possible to avoid backtracking completely, it was decided to limit its use as much as possible. This was because every time a strategy is assessed, an evaluation of the economics model is made. To avoid excessive computation times, the number of times the economics model is evaluated should be kept low if possible. When backtracking was used, it had to be not only limited, but also predictable in extent. This system was designed for use in an industrial environment, and the test strategy planning process should not take more than a few hours in automatic mode. The time given by industrial users as 'reasonable' was 2-3 hours, although longer running times were acceptable for very large, complex designs, where the benefits of test strategy planning are proportionately greater.

Finally, it was noted that the TDES/TIGER system evaluated TDMs only in local terms. It was largely assumed that, once the relevant I-paths were established, TDMs were independent of each other, with the exception of the 'shareability' measure. No active optimisation of the whole design was considered. The ECOtest philosophy, with the use of an economics model for the whole design, takes a more global view. Ideally, all interactions between test methods should be considered in order to arrive at a maximally optimised solution. Although this goal is difficult to achieve completely, it was kept in mind during the development of the test strategy planning algorithms.

4.6.2 Creating solutions to the test strategy planning problem

In order to evaluate the effectiveness of test strategy planning algorithms and their run times compared to the exhaustive solution set, the total solution space has to be quantified. This can be done if the allocation of design for test methods to functional blocks in a design is described as a tree. The root of the tree is the overall design. Each node in the design is a leaf node, with branches representing the test methods applicable to that node, and a cost associated with each branch. Each test method then becomes a node and branches out to the number of the remaining blocks in the design. An example tree is shown in figure 4.4, where the rectangular blocks represent circuit blocks and the filled triangular markers represent the test methods evaluated.

The total number of solutions in this case would be:

$$n_{ts} = n_{tu}! \times \prod_{i=1}^{n_{tu}} n_{tm}(i)$$

where n_{ts} is the number of possible test strategies, n_{tu} is the number of functional blocks in the design, and $n_{tm}(i)$ is the number of test methods applicable to functional block (i). Each test method is assumed to fulfil the testability/accessibility objective and is the combination of a testability and an accessibility enhancing method for block i.

However, in this case, many solutions would be duplicated. The cost of a particular solution (functional block / test method combination) is the same regardless of the order in which the functional blocks appear in the solution. For example, in a three-block design, the solution [block1->testmeth_x, block2->testmeth_y, block3->testmeth_z] is equivalent to [block2->testmeth_y, block3->testmeth_z, block1->testmeth_x]. Therefore, only a sub-tree needs to be evaluated to provide an exhaustive solution set. Figure 4.5 shows the solution tree that would provide the solution set.

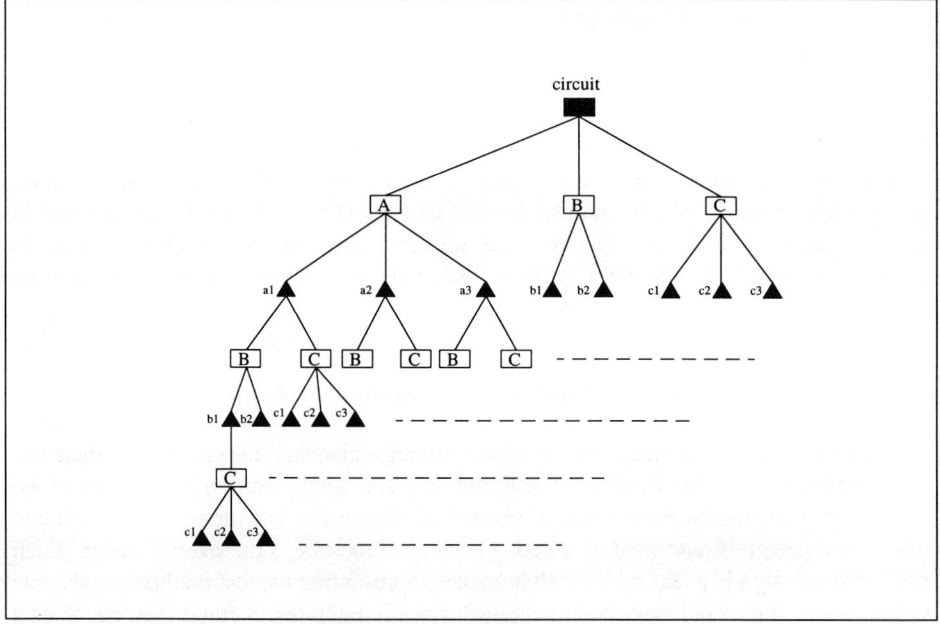

Figure 4.4 Test strategy planning solution tree

The unfilled triangular markers represent solutions that do not need to be evaluated. The total number of different solutions is:

$$n_{ts} = \prod_{i=1}^{n_{tu}} n_{tm}(i)$$

In practice, the number of possible solutions rapidly becomes too large for an exhaustive search. From the above equation, if we consider the case of a circuit consisting of ten blocks with eight test methods applicable to each block, the total number of solutions is 8^{10} = 1 073 741 824. Assuming the evaluation of each strategy takes at least one second, the time taken to calculate all the solutions would be around 34 years!

It is obvious that only a subset of the solution space can practically be explored, and this needs to be chosen in a way that will achieve a minimum (or near-minimum) cost solution.

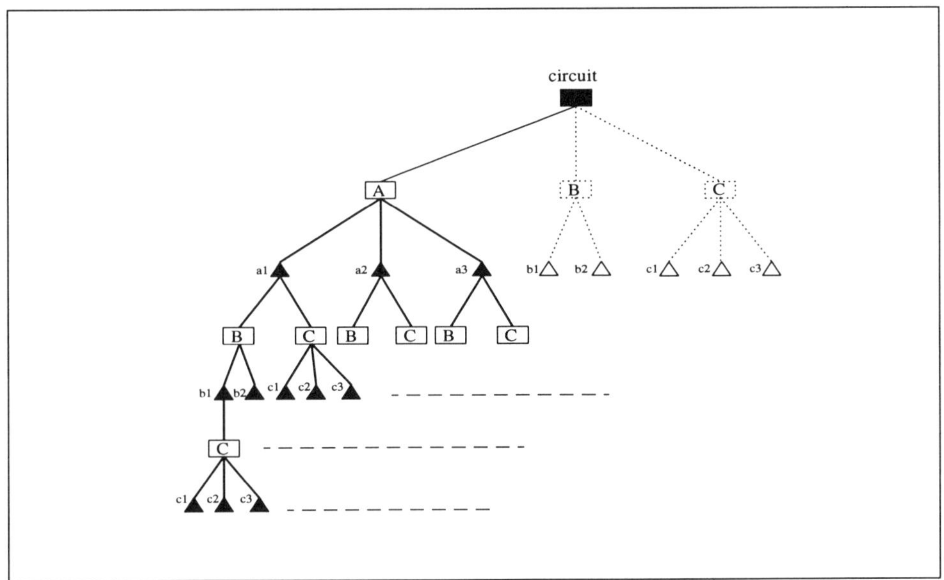

Figure 4.5 Exhaustive solution tree

Once a solution is finalised, the order of evaluation of test methods is not important. However, during evaluation, a test method applied to a block can influence the cost effectiveness of test methods for subsequent blocks. For example, the cost modelling of test application costs assumes that an initial number of test patterns is applied as part of the production price. Extra test patterns incur a charge which is fixed within a charge band related to the pin memory size of the ATE used. This can make the cost effectiveness of the method very sensitive to the number of test patterns in the boundary between charge bands. It is possible, during a non-exhaustive search, to reach a local minimum in the cost function. This will not happen in every design; its occurrence depends on the types of functional block present, the economics model parameter values, and the test methods considered. Although a local minimum will still offer a better solution than the starting point, it may cause a potentially better solution to be missed by the system, and the goal of a test strategy planning method is to achieve a global cost minimum. However, a local minimum may in fact be within a small cost interval of a global minimum, and a decision has to be made on the trade-off between global cost effectiveness and effort and time searching the solution space. This decision will be influenced by the user's expectations from the system. If the system is used as a guide in the early stages of the design process (and this is where it will be at its most useful), then the accuracy of the parameters passed to the economics model is not normally great enough to allow decisions based on minor cost differences. Achieving a global minimum or near-minimum solution becomes more important when the system is used at a stage where accurate parameter values are available and the user is not likely to improve on the system's solution at a later stage.

The test strategy planning process can be made simpler if the problem is tackled in two stages: testability improvement for individual blocks, and accessibility improvement for all the circuit bundles. It has already been mentioned that test methods in the data base are classified into testability and accessibility enhancing. The two categories are not mutually exclusive, as some testability enhancing methods may also improve a block's accessibility. Nevertheless, this differentiation helps to separate test strategy planning into distinct stages. This is essential in order to be compatible with the economics model definition. When costs are calculated by the economics model, accessibility to the blocks that make up the design is assumed. That is, each i/o line of a block is both controllable and observable. The system warns the user if this objective cannot be fulfilled by the system. The experience of the Alvey model, which included an accessibility value of between 0 and 1 for each block as a primary parameter, for adjusting the costs of test pattern generation and cost application, showed that it was extremely difficult to estimate accessibility values reliably and determine the fashion in which the costs would be adjusted as a result of accessibility changes.

By dividing the algorithm into stages, a logical progression is followed; i.e., major changes (testability enhancement) are dealt with first, thus solving a large part of the problem, and probably resulting in accessibility improvement. The second part of the algorithm deals with the optimum way of providing accessibility where necessary. Thus, unnecessary DFT structure placement can be avoided. The alternative is to provide accessibility for the circuit first (using the accessibility methods), and then tackle the testability objective. This leaves the problem of removing unnecessary DFT structures as the test strategy planning progresses.

Either way, the creation of two separate objectives has the effect of reducing the number of different solutions possible. Test methods can be divided into testability and accessibility methods. For example, four testability and two accessibility enhancing methods would result in eight unique test method combinations. This was the number considered in the previous example, where the objective was simply to allocate test methods to blocks. If the test strategy planning is done in two stages, and the process is split into satisfying testability and accessibility objectives, then the solution tree is divided and the number of different solutions is:

$$n_{ts} = \prod_{i=1}^{n_{tu}} n_{tmtest}(i) + \prod_{i=1}^{n_{tu}} n_{tmacc}(i)$$

where $n_{tmtest}(i)$ is the number of testability enhancing methods applicable to block i and $n_{tmacc}(i)$ the number of accessibility enhancing methods applicable to the same block. Using the previous ten block design example, if four testability enhancing and two accessibility enhancing methods were applicable to each block, the total number of solutions would be $4^{10} + 2^{10} = 1\ 049\ 600$, and at a rate of 1 second per solution evaluation would take around 12 days to calculate (as opposed to the 34 years quoted earlier). Although this division of goals introduces the possibility of missing a global minimum solution, in reality its use is justified in terms of ease of computation and cost model calculations. The cost of accessibility improvement is small compared to the savings made by applying cost effective testability enhancing methods, and any deviation from a global minimum is likely to be small. A number of different test strategy planning algorithms have been implemented and they are described below. A circuit example will be used to illustrate some of the points discussed.

4.6.2.1 An example circuit

Figure 4.6 shows an example circuit based on an industrial design, which will be useful in illustrating the test strategy planning process. In the example, node 1 is a PLA, nodes 2 and 3 are random sequential logic, and node 4 is a RAM.

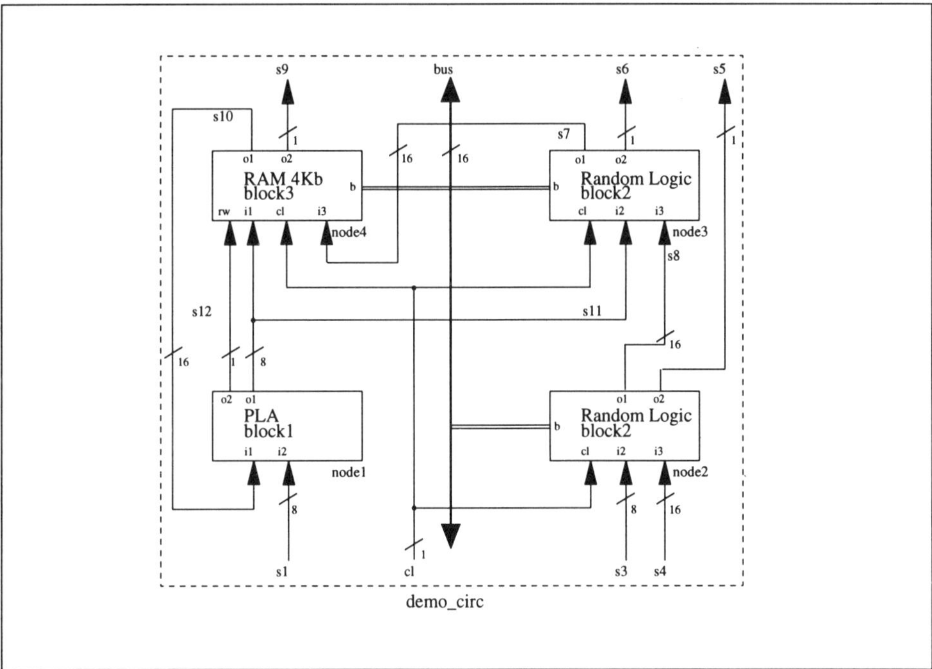

Figure 4.6 A circuit example

The starting strategy for the system had to be as close to 'no design for test' as possible. However, for the system to be initialised, this state had to be defined. This was simply a case of setting default starting points, and does not influence the effectiveness of the test strategy planning. The default starting strategy for the PLA was an exhaustive test set, which results in very high test application costs. This may not be a realistic choice for PLAs with a large number of i/o lines, but it is an easily defined state and, as mentioned before, should not reduce the ability of the test strategy planner to achieve a good solution. 'No design for test' was chosen for the random sequential blocks, i.e. an automatic test pattern generator in combination with manual test pattern generation was to be used. A standard March test was chosen as the starting RAM strategy. No accessibility enhancing methods were used in the initial state.

4.6.3 Exhaustive test strategy evaluation

Although, in practical terms, an exhaustive test method selection process would not be acceptable due to the long run times and memory requirements, a set of test strategies was evaluated exhaustively for the example circuit demo_circ. This was done to determine the success or otherwise of the test strategy planning algorithms and to examine trends arising from the economics modelling used, thus providing a useful baseline.

Some methods, already known to be expensive for the particular circuit (such as the exhaustive test method for the PLA) were discounted before the search began in order to save on computation time, and the set of test methods included only three testability enhancing and three accessibility enhancing methods for each block, thus giving nine possible test method combinations in each case, and a total of 6187 possible solutions. A simple recursive algorithm was used to evaluate the solution tree, using pre-order traversal. At each node of the tree, the left hand sub-tree was evaluated first, until a leaf node was reached, at which point the program would return up the tree to the next available sub-tree, which would be evaluated in the same way. Only three test methods per block were used. These were chosen as the three most cost effective methods for the block when the test methods for the other blocks were set to their default values. The methods chosen in this way included all of the methods that appeared in the solutions provided by all of the test strategy planning algorithms. As only two test methods for each block appeared in the solution in this case, a third method was included as a safety margin. The test strategies were evaluated for a production volume of 200 000 units, as this was a value for which the algorithms would occasionally only reach a local minimum. The methods selected were as shown in table 4.5.

The exhaustive algorithm included a check on the accessibility status of the circuit after a test strategy was selected, as not all the test strategies evaluated would satisfy the accessibility objectives. The ones that did not were rejected as illegal states, leaving a total of 5108 valid strategies. The minimum and the maximum costs together with the related test strategies are shown in table 4.6.

Table 4.5 Test methods used in the exhaustive evaluation

Block type	Testability methods	Accessibility methods	Method no.
Block 1 (PLA)	treuer	ext_nodft	1
	fujiwara	ext_scan_set	2
	hfc	ext_d_latch	3
Block 2 (RS)	int_nodft	ext_nodft	1
	int_scan	ext_scan_set	2
	int_scan_set	ext_d_latch	3
Block 3 (RAM)	march	ext_nodft	1
	march_self	ext_scan_set	2
	illman	ext_d_latch	3

Table 4.6 Most and least cost effective strategies for 200K production volume

Cost (DM)	Node 1		Node 2		Node 3		Node 4	
	test.	acc.	test.	acc.	test.	acc.	test.	acc.
5 688 325 (min)	1	1	1	3	2	1	3	1
	1	1	2	1	1	3	3	1
8 755 663 (max)	2	2	1	2	3	2	1	3
	2	2	3	2	1	2	1	3

There are two strategies for each case, as they are in fact equivalent. Nodes 2 and 3 are instantiations of the same block, and in the above strategies the methods are simply switched round. The average test strategy cost was 7 040 616 DM. It is clear from these results that although the test methods selected were already a cost effective set, there is still a 35% saving if the minimum cost strategy is selected rather than the maximum. In real terms, of course, the saving depends on the default test methods chosen.

The test strategies were ordered in terms of cost, and the resulting curve is shown in figure 4.7. The change in cost shows a marked stepwise reduction. This is in fact due to test vector application costs, and arises from the pricing function, shown in figure 4.8. The test pattern application cost modelling is based on vendor pricing information. In this implementation of the model, the production price included the application of 32K of test vectors. Each additional 32K (or less) incurred a cost of 5 DM per IC for test pattern application. The test vector step is usually linked to the pin memory of the ATE used. Each extra loading of test patterns into the memory effectively incurs an additional cost.

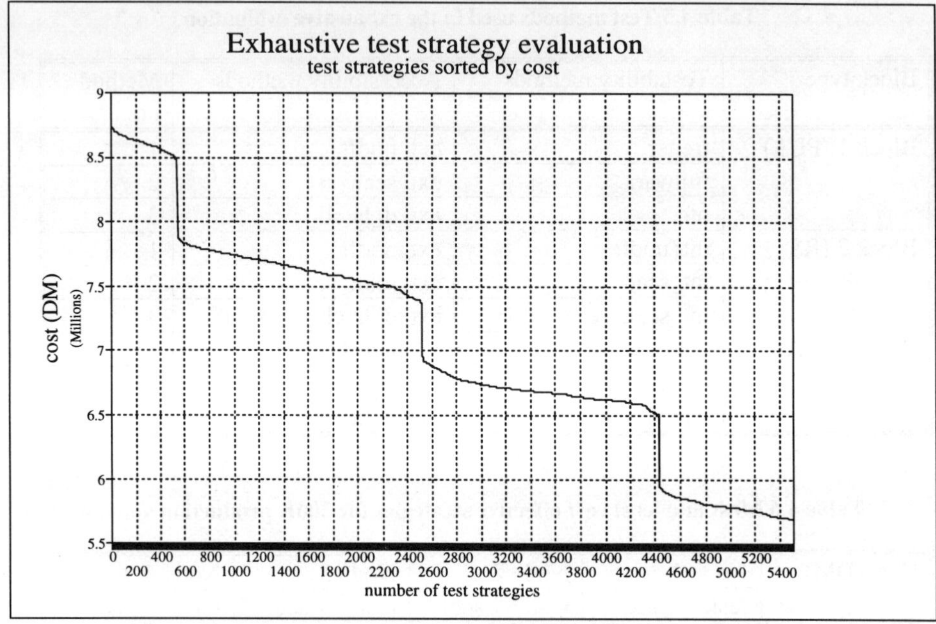

Figure 4.7 Exhaustive test strategy set sorted by cost

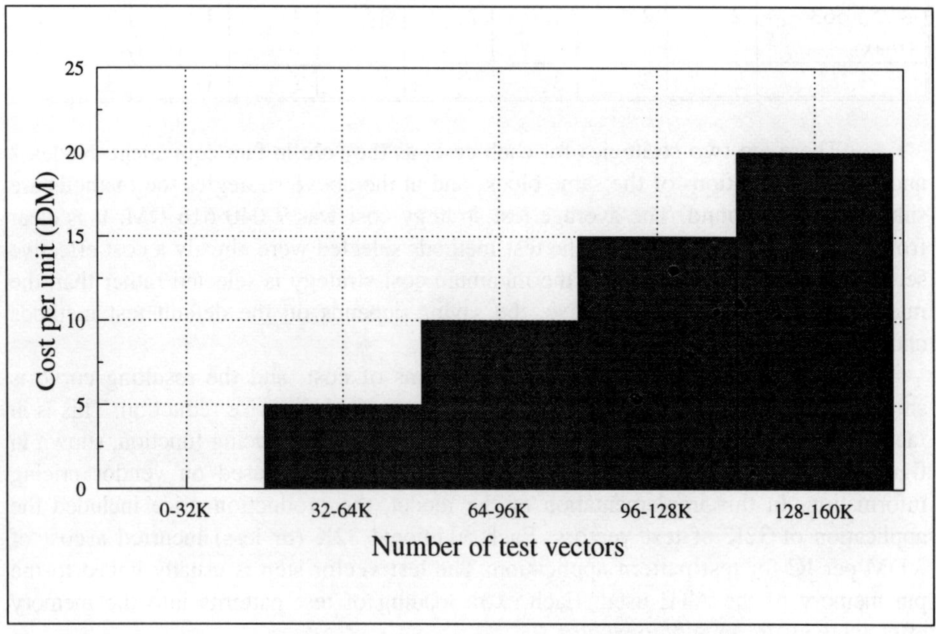

Figure 4.8 Test pattern application pricing function

This pricing policy means that the number of test patterns can be the only determining factor in test strategy selection at the border between price bands, especially at high production volumes. A small change in the number of test vectors may push the test pattern application cost into the next (higher) pricing band. The steps in the test strategies graph occur in exactly this kind of situation.

This effect is obviously entirely dependent on the cost modelling used and introduces an element of error if the test vector calculation by the cost model is not exact. In this case the cost model was designed in-house, and the effect was known. In some cases, it may be better to use a linear pricing function in the early stages of the design if the number of test vectors to be used is not known or cannot be accurately calculated.

However, the number of test vectors is by no means the only factor that will determine a test strategy. Within each pricing band, considerable savings can be made (up to around 0.5 million DM in this case). These savings will depend on the other economic factors, as the test application cost remains constant within each band.

4.6.4 Test strategy planning: application of testability and accessibility enhancing methods

The objective of the test strategy planning algorithms is twofold: firstly, to solve the testability problem for each individual testable unit (TU) in the design, and secondly, to ensure the accessibility of every TU i/o line, so that test patterns can be propagated to the block and the responses observed. Figure 4.9 shows the Nassi-Schneiderman chart of the algorithm (algorithm 1). This is auto_plan1 in the test planner commands.

The testability objective is tackled first, as it will often create the largest economic impact, and may also solve, at least partly, the accessibility problem. Accessibility enhancing test methods can then be applied only where needed. The basis of the algorithm is one of iterative improvement; a solution is evaluated, and accepted if it offers a cost reduction, as calculated by the economics model. The use of the economics model takes into account the status of the whole design at one time, and ensures that the evaluated solutions reflect costs in the whole of the circuit and not only the block under consideration. However, the iterative improvement action has to be guided so that excessive iterations are avoided, and an optimal solution can be reached with a relatively short computation time. The classification of objectives into testability and accessibility ones helps in this, as do the initial checks made on the suitability of a test method to a block (in terms of design type and style) before the cost model is evaluated.

The testability enhancement section selects a block (called a testable unit), applies all suitable methods to that TU and selects the cheapest. That method is then fixed, and the next block is selected. At the first implementation, the ordering was taken from the circuit description, and no attempt was made to order the blocks for test method application. At the end of the pass, all blocks should have a testability method assigned to them that is equally or more cost effective than the starting solution. The accessibility status of the circuit is also likely to have changed, as some testability method structures can also be used for accessibility enhancement. For example, the Illman method [Illman86] for RAM testing is a self-test testability enhancing method which also provides accessibility improvement to the RAM, as it uses LFSRs which can be utilised for accessibility improvement.

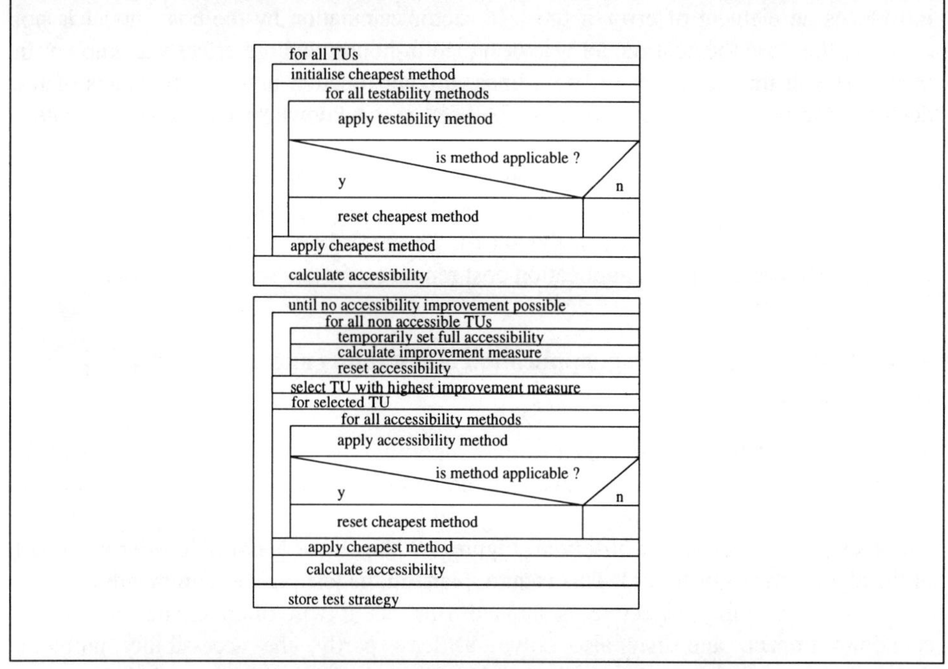

Figure 4.9 Test strategy planning algorithm 1 chart

At the next stage of the algorithm, the accessibility objective has to be satisfied. For this, the current accessibility status of each block is evaluated. At this point an addition to the basic algorithm is made. There may be several ways to improve the accessibility status of the circuit as a whole, and this should be done in the most economical way. It is therefore important that the minimum number of changes are made to the circuit. As mentioned before, external test methods are shown to incur a financial penalty, due to the assumptions made in the cost model. Although they are necessary, the system should ensure that their application is optimised.

In order to achieve optimisation of external test method application, the blocks where an accessibility improvement would create the largest overall improvement in the design were targeted first. For example, a control block which provides signals that control transparent paths in other parts of the design, may be a good choice for external test method application, depending on circuit structure. To determine the most critical block in the design in terms of accessibility improvement, each non-accessible block was temporarily made accessible (i.e. its inputs and outputs were set to pseudo-i/o). Then the accessibility analysis algorithm was run, in order to calculate the number of bundles (and therefore the number of signal lines) that were made accessible as a result. The number of extra lines made accessible forms the improvement measure. The accessibility status of all bundles was reset to its previous value before the next non-accessible block was tackled. Once the block which will create the maximum global accessibility improvement is determined, all suitable external methods are applied and the cheapest chosen for the block. External test methods are applied to a block in an optimised fashion, i.e. the test method is applied only to the non-accessible bundles of a block. This is to avoid unnecessary application of test methods which would reduce the cost benefits.

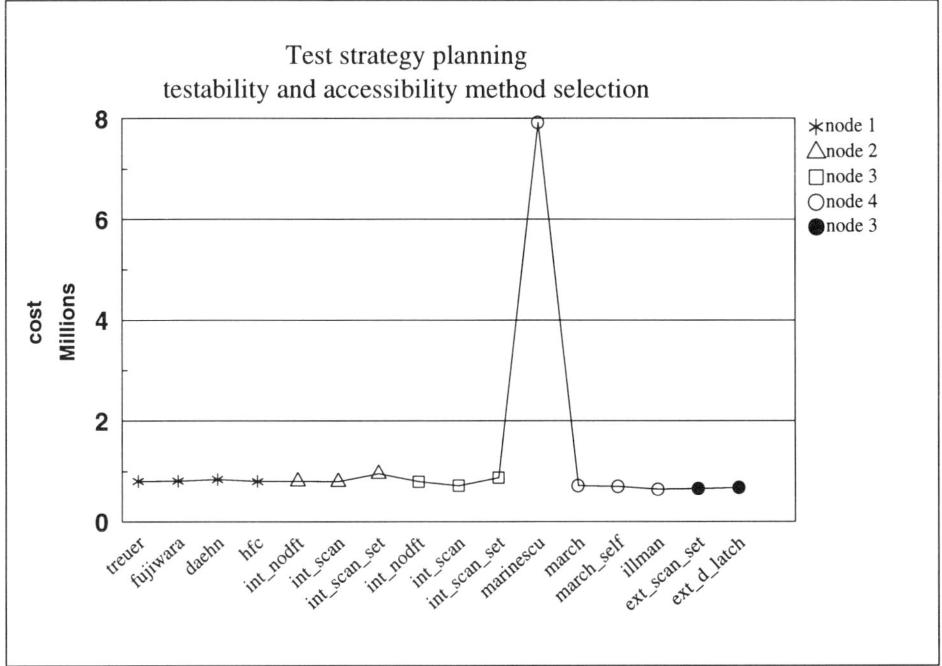

Figure 4.10 Algorithm 1: costs at each evaluation

Figures 4.10 and 4.11 show the costs at each stage. Figure 4.11 is simply figure 4.10 plotted on a different scale to show the test methods chosen and their costs. The vertical lines in figure 4.11 denote the different stages of test method application, and the test method chosen at each stage is underlined.

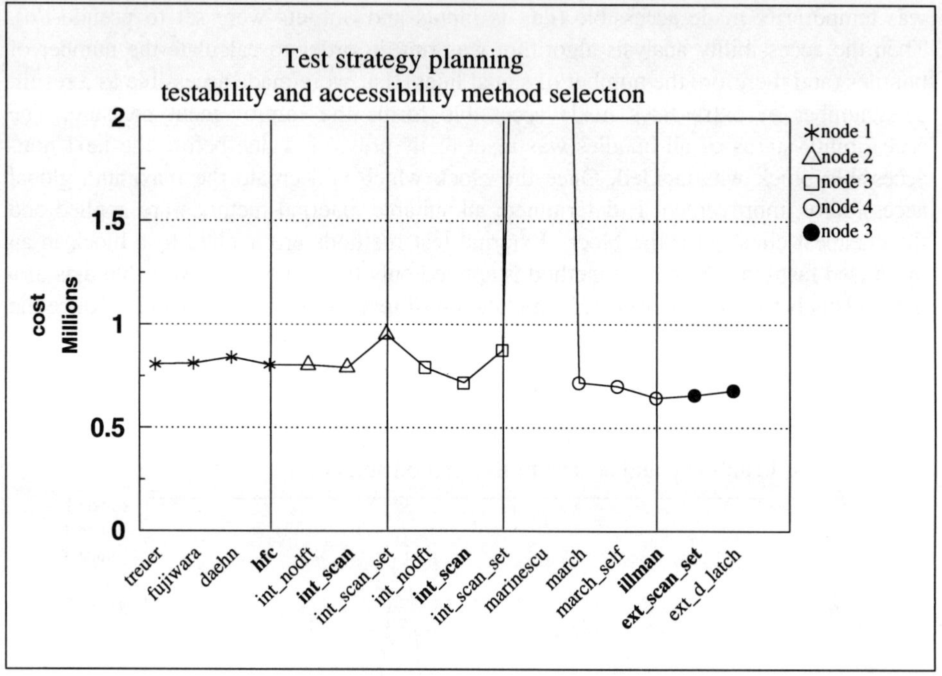

Figure 4.11 Algorithm 1: costs at each evaluation (detail)

Figure 4.12 shows the path to the solution in the example circuit in terms of the solution tree. At every stage of the application of testability methods, test methods are effectively ordered in terms of cost and the cheapest path chosen. At the accessibility test method application stage, the blocks in the design are ordered in terms of their accessibility, before the cheapest one is chosen.

The advantages of this algorithm are that is simple to implement and a cost effective solution is achieved with a low number of cost model evaluations and accessibility analyses. In this case, the computation time is of linear complexity and is approximately

$$t_{tsp} = t_{cm} \times \sum_{i=1}^{n_{tu}} n_{tmtest}(i) + (t_{cm} + t_{accan}) \times \sum_{j=1}^{nonacctu} n_{tmacc}(j)$$

where t_{tsp} is the total computation time for test strategy planning, t_{cm} is the time taken for a cost model evaluation, and t_{accann} is the time taken for accessibility analysis. *nonacctu* is the number of non-accessible testable units left. This number is adjusted after an accessibility method has been chosen for a TU, as this may have an impact on the accessibility of other TUs. The maximum value of *nonacctu* is the number of TUs in the design.

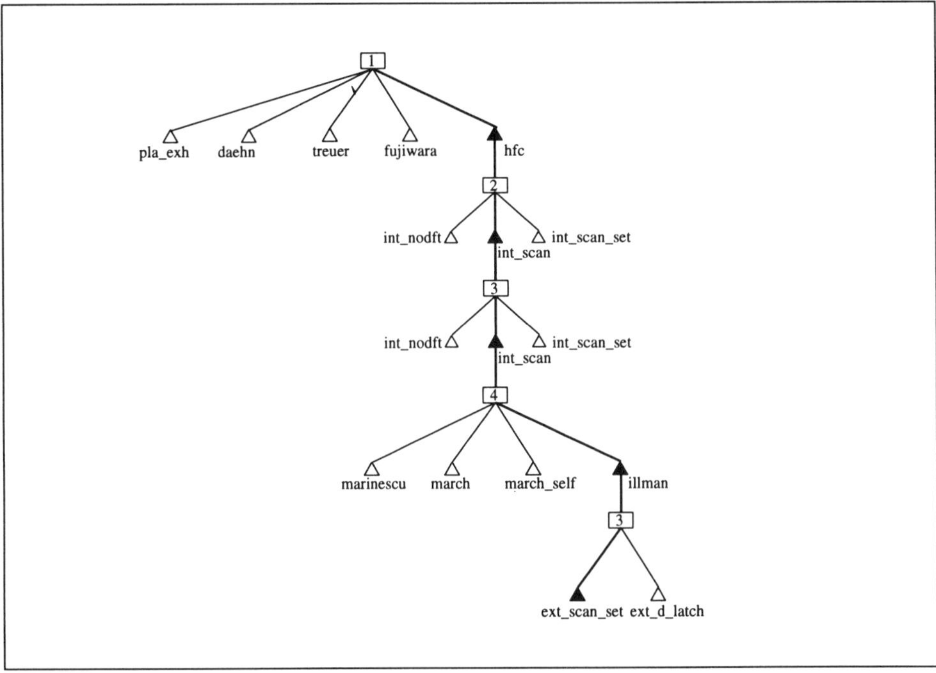

Figure 4.12 Solution path of algorithm 1

However, it is possible that the most cost effective solution calculated by this method is only a local minimum. Whether or not that happens depends not only on the design but also on the economics model parameters, such as the production volume, which can alter the relative importance of test method effects. The order of evaluation of TUs also matters. In this case, the TUs are not ordered for testability method application. For example, for a production volume of 200 000 ICs, the test strategy shown in table 4.7 is obtained using the above algorithm.

Table 4.7 Test strategy for 200 000 ICs, using algorithm 1

TU	Testability method	Accessibility method
Node 1	treuer	ext_nodft
Node 2	int_nodft	ext_nodft
Node 3	int_nodft	ext_d_latch
Node 4	illman	ext_nodft
Cost = 6 606 879 DM		

In fact, a more cost effective solution exists, which can be achieved by applying int_scan to node 3, giving a cost of 5 688 801 DM, a saving of almost 14%, although, at the initial volume of 10 000 units, the algorithm achieved the global minimum. Some solutions to the local minima problem are discussed in the following sections.

4.6.5 Evaluating multiple branches

One way to increase the probability of achieving a global minimum is to increase the coverage of the solution tree beyond the 'cheapest path' option adopted by the previous algorithm. However, the number of solutions evaluated has to be controlled so that excessive run times are avoided. One option is to accept the two cheapest solutions at each stage of test method application. At the next test method selection, four solutions will be selected. Of these, the two cheapest are accepted and form the two new paths. In that way, two solution paths are set up and followed. Evaluated solutions are placed in the temporary solution list before the two cheapest to be placed in the solution paths are chosen. The selection and evaluation of TUs proceeds sequentially, and there is no backtracking involved. The computation time is almost twice that of the previous algorithm, and is given by

$$t_{\mathrm{sp}} = 2 \times (t_{\mathrm{cm}} \times \sum_{i=1}^{n_{\mathrm{tu}}} n_{\mathrm{tmtest}}(i) + (t_{\mathrm{cm}} + t_{\mathrm{accan}}) \times \sum_{j=1}^{nonacctu} n_{tmacc}(j)) - t_{\mathrm{cm}} \times n_{\mathrm{tmtest}}(1)$$

Although the chances of achieving a global minimum are better due to the fact that a larger number of solutions are evaluated, the final solution is sensitive to the starting TU. Although it is possible that the two solution paths can 'cross over', i.e. the path that started out more expensive can become the more cost effective of the two, there is always the possibility that the most cost effective path overall can still be missed. For example, applying the algorithm to demo_circ for a volume of 200 000 and using the same order of TUs as in the previous example, the final solution is no different from the one achieved using algorithm 1 (described in the previous section).

4.6.6 Ordering the TU processing

Although the approach of algorithm 1 provides a cost effective solution, there is little guidance in the application of the testability enhancing methods, and as a result it is possible to encounter a local minimum before the global minimum is reached. It is certainly possible, due to the lack of any ordering of TUs before test method application, to tackle TUs with minimal influence on other blocks or minimal testability problems first, thus leaving less scope for test method choice for other, more critical TUs. Many users, when attempting manual test strategy planning, would find the most cost effective testability method for blocks which are either critical to the function of the circuit (e.g. a control PLA) or difficult to test, before attempting to apply testability methods to the rest of the circuit. The approach is analogous to that adopted in algorithm 1 for accessibility test method application: critical TUs are tackled first.

A method is therefore required for recognising critical TUs prior to the application of testability enhancing methods. At the level of abstraction at which the test strategy planner is used, there is seldom enough information to use a testability analysis package such as CAMELOT [Bennetts80]. However, there is information available on the accessibility of bundles and on transparency relationships. It is possible to order TUs on the basis of an accessibility improvement measure, as used in algorithm 1 for accessibility test method selection. This does not amount to a full testability analysis, as there is no guarantee that the testability method chosen for a non-accessible TU will solve the accessibility problem, but it is used as a heuristic measure to give an indication as to how critical the TU in question is to the rest of the circuit.

In terms of the economics model, a larger, more complex TU is likely to have more bundles associated with it, and therefore more potential for accessibility improvement. Large TUs are also likely to have greater impact on area overhead and test application costs. Therefore both the size and complexity of the circuit are indirectly taken into account, both locally and globally, including any transparent paths. Other measures can also be used to guide the search, such as the size of the circuit, test pattern generation cost, etc. However, these only consider the TU itself with no link to the rest of the design.

Algorithm 2 (auto_plan2 in the test strategy planner) is based on ordering in accessibility terms. To minimise CPU time, TUs are sorted in order of decreasing improvement measure once, at the start of the program. The TUs are then targeted in sequence for the accessibility method selection. The approximate time taken for test strategy selection using this method is

$$t_{tsp} = t_{accan} \times nonacctu + t_{cm} \times \sum_{i=1}^{ntu} n_{tmtest}(i) + (t_{cm} + t_{accan}) \times \sum_{j=1}^{nonacctu} n_{tmacc}(j)$$

The only excess time element is that of ordering the TUs. In the example circuit, for 200 000 ICs production volume, table 4.8 shows the test strategy together with the order the nodes were processed for the testability methods and the accessibility improvement measure for each node. Accessibility method application was as in algorithm 1.

Table 4.8 Test strategy for 200 000 ICs, algorithm 2

TU	Access improvement	Testability method	Accessibility method
Node 4	48	illman	ext_nodft
Node 3	41	int_scan	ext_d_latch
Node 1	16	treuer	ext_nodft
Node 2	16	int_nodft	ext_nodft
cost = 5 688 801 DM			

Although the ordering in algorithm 2 is only performed once, it is possible, if the extra computation time is not critical, to reorder the TUs at every stage of the testability method selection. The user has to determine the trade-off between the value of the heuristic measure and the extra computation time involved. In this case, if ordering of the nodes is done at each stage of the testability method selection, the order changes to node 4, node 1, node 3, node 2. The final plan is not altered, but the change in sequence might be important in different circuits. The method of accessibility ordering is most useful in circuits where transparency relationships are known, and accessibility effects can be reliably propagated.

4.6.7 An iterative solution

All of the above methods avoid backtracking. However, it may be necessary to backtrack in order to move away from a local minimum. As there is no way a local minimum can be recognised as such until a cheaper solution is found, the decision as to where to backtrack to is a difficult one. Additionally, any backtracking should be predictable in terms of the time it takes. A compromise was reached by adopting an iterative solution, which is essentially one for improving a local minimum. Instead of exploring two separate paths, the test strategy selection process is run through once, and then the nodes are processed again by running the algorithm once more with the strategy from the previous run as input. The running time is only slightly longer than the two-branch algorithm (all nodes are processed twice), but the technique is powerful in removing the sensitivity to the order of TU processing, and has become the preferred method for test strategy planning in ECOtest. In the example circuit, this arrives at the solution shown in table 4.9.

Table 4.9 Test strategy for 200000 ICs, using iteration

TU	Testability method	Accessibility method
node 1	treuer	ext_nodft
node 2	int_scan	ext_nodft
node 3	int_nodft	ext_d_latch
node 4	illman	ext_nodft
cost = 5 688 325 DM		

The cost is in fact 0.008% lower than the solution arrived at by the accessibility ordering method, and the saving is not significant. In design terms, the two solutions are very similar, and this is reflected in the similar cost. This is one example where a local minimum is so near to a global minimum that is forms an equally good solution. The differentiation between two similarly priced test methods is as accurate as the modelling and the parameter values in the cost model allow.

4.6.7.1 Relative costs of test strategies

Using the preferred iterative algorithm, the following strategy was obtained for a production volume of 20 000 units for demo_circ (table 4.10). All the available methods for each functional block were included in the search.

Table 4.10 Optimum strategy for 20000 units

Functional block	Testability method	Accessibility method
Node 1 (PLA)	hfc	ext_nodft
Node 2 (random sequential)	int_scan	ext_nodft
Node 3 (random sequential)	int_scan	ext_scan_set
Node 4 (RAM)	illman	ext_nodft

Figure 4.13 shows the cost progression as each of the test methods is applied. It can be seen from the graph that design costs form a significant part of the final cost. The development costs are only incurred once and are more significant at low rather than high production volumes. There is an increase in production costs as more DFT methods are chosen. The manual test pattern generation cost element is reduced to nil as more DFT methods are used, which makes it possible to rely on automatic test pattern generation alone. The price bands for test vector application are also seen here; the choice test strategy moves the cost from the third pricing band (10 DM per IC, over 64K of test vectors), to the first band (less than 32K for test vectors, no extra cost). In fact, in the default test strategy, the test pattern application cost was even higher, as an exhaustive test set was assumed for the PLA. The number of test vectors is reduced due to the use of a self-test method for the PLA, and scan for the random sequential blocks which effectively reduces the sequential depth. Finally, self-test was used for the memory, reducing the number of test vectors to less than 32K.

The final test method applied is an accessibility enhancing one. The accessibility analysis at the end of the testability enhancing test method application showed an accessibility problem for bundle s_8, and external scan was used to remedy this. This in fact incurred an extra cost, which can be viewed as a correction factor for the cost model, as extra test costs due to low accessibility in parts of the circuit are not modelled explicitly in the cost model.

The same exercise was repeated for a production volume of 200 000 units. The costs at each evaluation stage for the resulting test strategy are shown in figure 4.14.

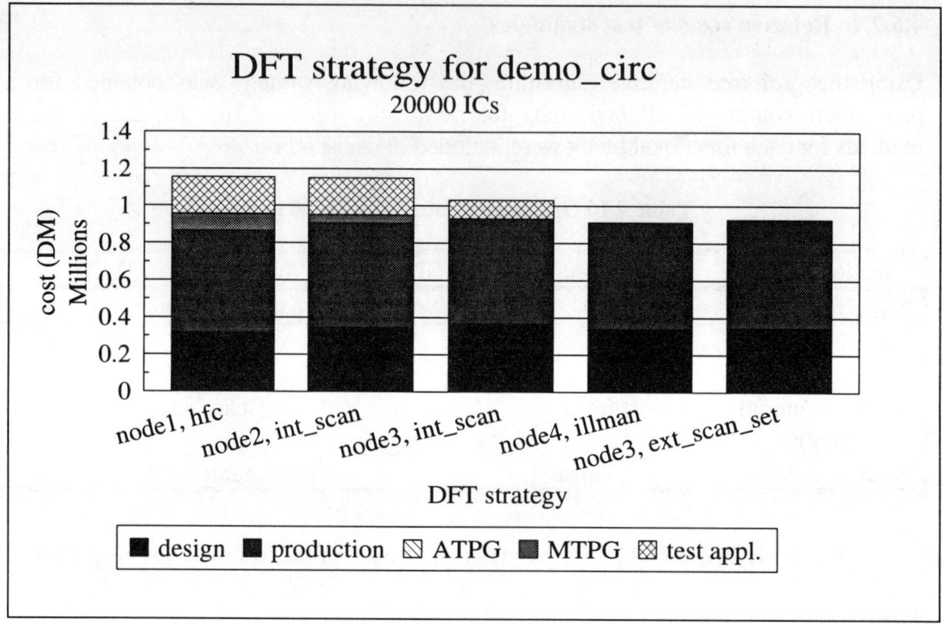

Figure 4.13 Test strategy cost progression, 20 000 ICs

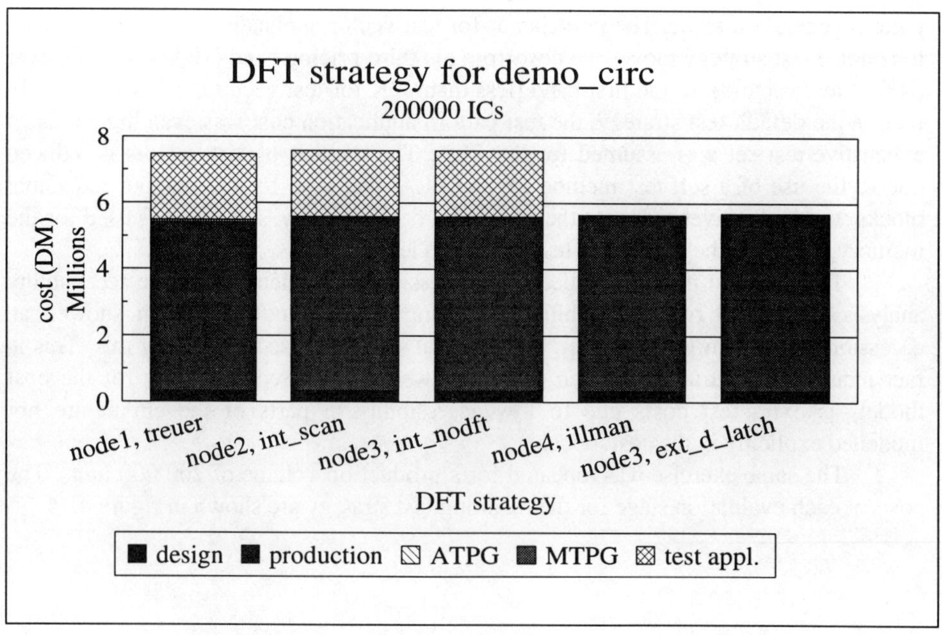

Figure 4.14 Test strategy cost progression, 200 000 ICs

As expected, non-recurring costs such as design and test pattern generation become less important for high volumes, while volume dependent costs such as production and test application now form a higher proportion of the total. Because of this, the test strategy changes towards a strategy which achieves a lower gate overhead even at the expense of greater test pattern generation costs. The most important change between the test strategies is that scan is no longer used for node 3. Sometimes using less DFT may be justified for higher volumes, providing testability and accessibility targets are met.

4.6.7.2 Processing a larger circuit

The example used for the above results was a fairly small circuit (four functional blocks and a total of 16 200 gates), but ECOtest can handle large, complex designs in the same way. The actual gate count of the circuit is less important than its connectivity complexity. A larger real-world design (called here demo_large) was evaluated using ECOtest. This consisted of ten blocks, two of which were RAMs and one a PLA. The rest of the blocks were random combinatorial or sequential, varying in gate count from 4000 to 45 000 gates, and of varying sequential depth. The total gate count was 257 547 gates. The circuit diagram of demo_large is shown in figure 4.15.

The optimal test strategy calculated by the system using the preferred iterative algorithm is shown in table 4.11.

Table 4.11 Test strategy for demo_large (for 20 000 ICs)

Block	Testability method	Accessibility method
N_1 (combinatorial random logic)	int_no_dft	ext_no_dft
N_2 (sequential random logic)	int_no_dft	ext_d_latch
N_3 (seq. random logic)	int_no_dft	ext_scan_set
N_4 (RAM)	march	ext_no_dft
N_5 (seq. random logic)	int_d_latch	ext_d_latch
N_6 (seq. random logic)	int_d_latch	ext_d_latch
N_7 (seq. random logic)	int_no_dft	ext_scan_set
N_8 (seq. random logic)	int_no_dft	ext_no_dft
N_9 (PLA)	hfc	ext_d_latch
N_{10} (RAM)	march	ext_no_dft

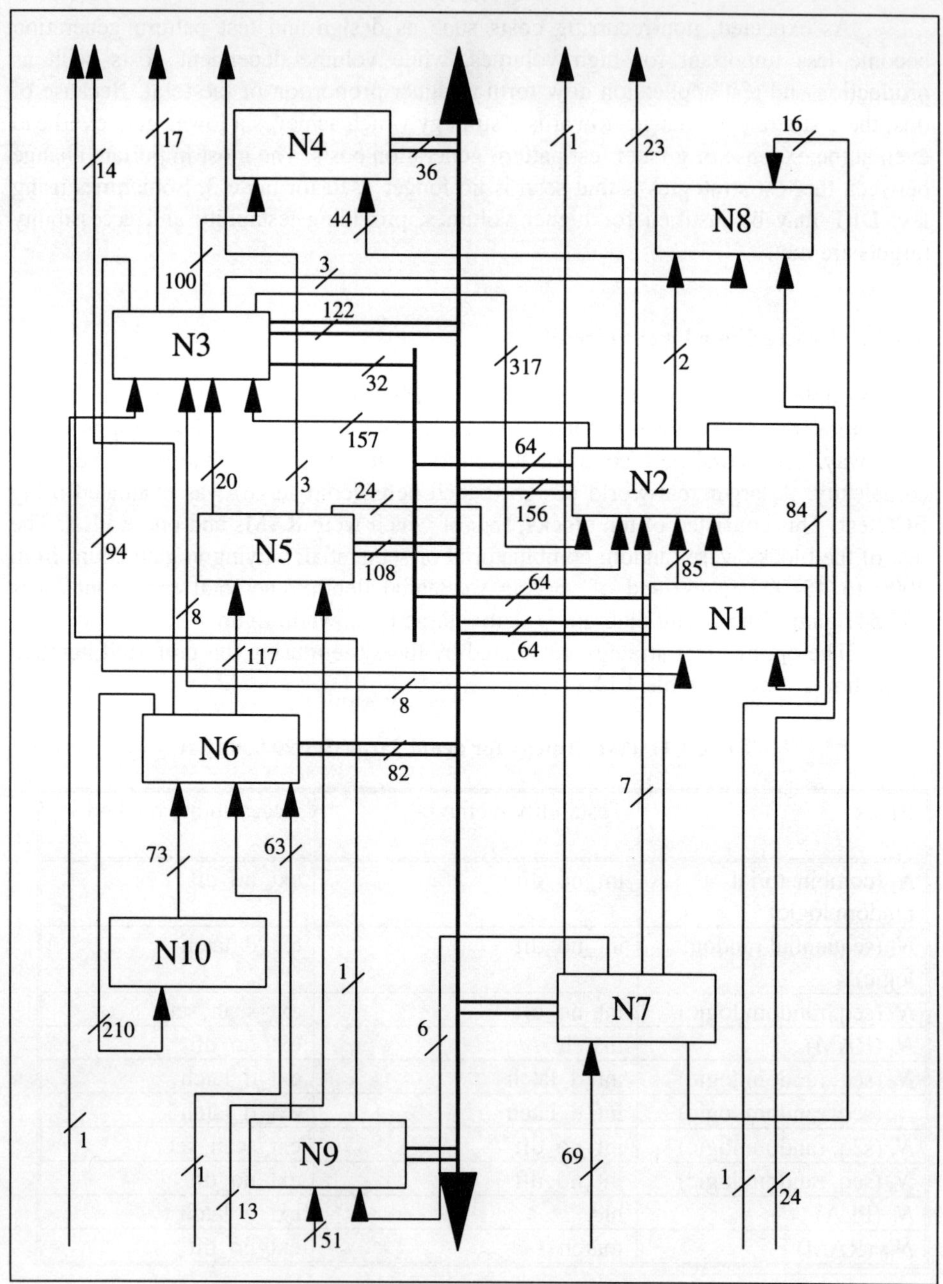

Figure 4.15 The demo_large circuit

For a volume of 200 000 ICs, the testability enhancing methods remained the same, but there were minor changes in the accessibility enhancing methods chosen. The final gate count was 272 126. Design for test methods were not chosen for every block. The two sequential blocks for which a testability method was suggested, were the ones of the greatest complexity, not only in terms of gates, but also in terms of sequential depth. Also, a different pricing function was used for test vector application, where an extra 5 DM per unit was charged for every 64K test vectors, effectively a cheaper option than the one provided in demo_circ. Under these circumstances, and bearing in mind that production costs and production overhead as a result of extra gates are high for a circuit of this size, a strategy where the minimum amount of design for test structures are implemented may be a cost effective option. The target fault cover specified by the user would still be achieved.

In order to provide a measure of the time taken to use the test strategy planner, some of the commands were timed for both circuits. The results are summarised in table 4.12. These CPU times can be reduced further by optimising the cost model evaluation.

Table 4.12 CPU times for test strategy planning functions

Command	demo_circ	demo_large
Setting up TSP	0:08.5	0:40.9
Apply 1 method	0:08.5	1:15.0
Accessibility status	0:21.6	3:12.6
Basic automatic planning algorithm	1:22.8	32:19.7

4.6.8 Further possible optimisation measures

So far, three different algorithms have been described: the basic fulfilment of testability and accessibility objectives, accessibility ordering, and iteration. Of these, the first uses the minimum of cost model and accessibility evaluations. The iterative solution takes approximately twice as long to achieve a solution, but would provide better results with most circuits. The basic algorithm would achieve a cost reduction, but may not approach a global minimum as closely as an iterative solution. The success of these and the accessibility ordering algorithms depends on the type of circuit, the different types of TU and test method available, as well as the default evaluation order. The basic algorithm will perform best where there is a large cost difference between strategies. A circuit with little TU variability such as the circuit demo_large, may be adequately covered by the basic algorithm. demo_circ was a relatively small circuit, but exhibited many local minima and therefore formed a good choice for algorithm comparison.

There is some further optimisation that can easily be performed using the test strategy planner functions. One situation that may arise is that a solution combines a testability and accessibility method for a particular TU. As testability and accessibility requirements are evaluated separately, it is possible that there may be a cost effective solution using a testability method that also improves accessibility, which may have been missed by the system. This situation can be optimised using a simple post-processing step of examining TUs which have both types of method assigned to them. It can then be seen if any method which combines testability and accessibility will further reduce the solution cost.

In the test strategy planner commands, three functions are offered which perform semi-automatic test strategy planning. The *int_methods* command will apply all internal (testability) methods to a TU and choose the most cost effective one. The *ext_methods* command does the same for the complete circuit, also evaluating where external methods are required. Finally, the *optimised_ext_methods* command does the final optimisation step described, and will replace an internal-external test method combination with an internal method that improves accessibility where it is cost effective to do so.

4.7 SIMULATED ANNEALING FOR TEST STRATEGY PLANNING

The algorithms discussed above will provide the user with a cost effective test strategy for implementation or manual improvement. However, they will not necessarily arrive at the globally optimum solution, and they do not have the option to adopt a more expensive local solution in the hope that it will lead to a global minimum. Although this would be acceptable in most cases, the user cannot tell how far from a global minimum the solution is. It was considered necessary to provide an alternative method of test strategy selection that had the ability to overcome local minima. The method chosen was simulated annealing, which is admittedly expensive in computation time. However, it adopts a different solution strategy, and can be a useful alternative to interactive or semi-interactive test strategy planning; the user can, for example, run the simulated annealing method overnight on a circuit that exhibits many local minima in order to achieve a solution which is optimal. Although a global minimum is still not guaranteed, there is a greater probability that it will be encountered or that the solution is close to it; how close it is depends on the parameters of the annealing algorithm. The annealing algorithm has the advantage that it can overcome local minima through a weighted random acceptance process. It has previously been used for VLSI design, and the best known application is for placement problems. However, it has not previously been applied to test strategy planning.

4.7.1 The generic simulated annealing algorithm

The simulated annealing algorithm is based on statistical mechanics, which is the study of complex systems made up of a large number of interacting atoms in thermal equilibrium at a finite temperature. Under equilibrium, the most probable states are those with the lowest energy [Wong88]. This is analogous to the combinatorial optimisation problem [Kirkpatrick83]. In physical systems, a stable configuration can be achieved by gradual lowering of the temperature and allowing the system to reach equilibrium at each temperature. Simulated annealing mimics this process. Placement and routeing problems in ICs, for which simulated annealing is often used to provide a solution, have analogies to the physical systems from which the algorithm was derived. The simplicity of the algorithm means that it can be readily adapted to a variety of combinatorial optimisation problems. Thus an algorithm of low complexity can be used for highly complex applications.

The simulated annealing algorithm can be used to achieve a 'minimum energy' (in our case minimum cost) state of the system. Because it incorporates a random element, the system should be able to escape local minima and work towards a global minimum, provided the control parameters are selected carefully. The generic algorithm takes the following form:

```
begin
 S := initial solution S0;
 T := initial temperature T0;
 while (stopping criterion is not satisfied) do
  begin
   while (not yet in equilibrium) do
    begin
     S' := some random neighbouring solution of S;
     D := C(S')-C(S);
     Prob := min(1, exp(-D/T));
     if random(0,1) <= prob then S := S';
    end;
   update T;
  end;
 output best solution;
end;
```

The algorithm starts with an initial solution and an initial (normally high) value of temperature (T). T is called 'temperature' because of the analogy to the physical annealing problem, but in simulated annealing it is simple a control parameter. At each temperature value, neighbouring solutions are evaluated and accepted or rejected until equilibrium is reached. The evaluation is done with a cost function $C(S)$ (for example, in a routeing problem the cost function may be the total wire length). If a cost reduction has been achieved, the solution is accepted. If not, it is given a probability of acceptance related to the control parameter T. At high T, solutions are more likely to be accepted, even if a cost reduction (D) has not been achieved. As T is reduced, this probability becomes smaller. A random element is introduced in the acceptance of solutions that do not reduce the cost function, which gives the opportunity to overcome local minima. Once equilibrium has been reached, T is reduced, and iteration continued until a stopping criterion satisfied. The stopping criterion may be simply related to computation time, or more likely to the improvement in the cost function achieved.

4.7.2 Implementing the algorithm in the ECOtest system

In the case of the test planner, the solution space is every combination of functional blocks and applicable test methods. An initial solution can be picked by randomly selecting a block, and then randomly selecting a test method from the set of test methods applicable to that block. A neighbouring solution is picked randomly, in the same way as the initial solution. The cost function $C(S)$ is the overall cost of the strategy, as calculated by the cost model. A neighbouring solution is always accepted if it is cheaper than the current one. If it is not, it is accepted with a probability of $e^{-D/T}$ where D is the cost difference from the previous solution and T is the current 'temperature'. Thus, some solutions which no not offer a cost improvement will sometimes be selected, allowing the algorithm to overcome local minima.

The initial annealing schedule was formulated as follows: assume that a 10% difference in cost is small enough for a more expensive solution to be considered. Then, at D = cost/10, the probability of acceptance would be around 50%. Using $Prob = e^{-D/T}$, this gives an initial T of (initial cost)/7. In fact it was found that 10% is a very generous limit, and 5% or less needed to be selected instead, based on the cost difference between neighbouring solutions. A 5% cost difference with a 50% probability of acceptance would mean an initial value of T of (initial cost)/70.

The system has to achieve equilibrium before T is updated (equivalent to lowering the temperature in the physical system). In the first instance, the algorithm was run through a fixed number of steps before the value of T was updated, and again there is a fixed number of updates. However, because of the nature of the solutions, there can be some very large decreases in cost as a result of a single method, for example by moving from an exhaustive test to a non-exhaustive one. In that case, if the cost drops by more than 20% from the starting value, then T is updated straight away and the next cycle begins.

The results of the algorithm showed that T set at cost/70 was a successful starting point. However, the fixed number of cycles presented a problem, as they were not an adequate check for equilibrium. The state of equilibrium is reached when either the maximum possible number of solutions have been accepted (making it unlikely that there is room for more improvement), or simply that little improvement is taking place. As a starting point, the cycle was run at each stage until the number of accepted solutions reached (n_{tu} - 1) * average_number_of_test_methods. This is the average number of neighbouring solutions in each case. If this condition is not fulfilled, it stops when twice that number of solutions have been attempted.

Finally, a termination condition needs to be present, to make sure that the process will not run indefinitely. The algorithm terminates when less than 10% of solutions have been accepted in the last run, excluding solutions of no cost difference. This indicates that little improvement is possible. The termination condition can be altered to allow more run time. To save on computation time, only solutions different from the existing ones for the TU were evaluated.

Some other considerations had to be taken into account in the application of test methods. If a solution fails to be accepted, it is important to reapply the previous solution before proceeding. This is not necessarily the previous solution applied by the simulated annealing algorithm, as that might concern a different TU, but the previous test method set for the TU currently processed. In order to achieve this, once a TU selection is made, the test methods that were applied to it are stored before a solution is attempted, so that they can be restored if necessary.

Another problem arises in that external methods can be applied to non-accessible blocks, and, in a later step, an accessibility enhancing internal method is applied to the same block, which obviously makes the external method redundant. In order to avoid unnecessary costs, internal methods are monitored for accessibility effects. If they also increase the TU's accessibility, ex_nodft is applied as part of the same solution. Note that the simulated annealing does not necessarily guarantee accessibility to all blocks, although it will achieve it in most cases. The final step in the process should be the same as in the other two test planning algorithms, which is to evaluate and fix accessibility problems in a cost optimal way.

4.7.3 A sample run

The control parameter T was set so that a 1% increase in strategy cost would have a 25% probability of acceptance. Each cycle was run until 24 strategies were accepted or a total of 48 strategies were attempted, whichever occurred first. Additionally, a cost reduction of more than 20% also resulted in the completion of a cycle, to cut down on computation time. The algorithm was terminated when less than 10% of attempted strategies with a non-zero cost difference (i.e. where the strategy was different to the previous one) were accepted in the last cycle. The values chosen for T reflected the fact that the difference between some local minima and the global minimum could be less than 1%, and also avoided oscillating too much between strategies. It still allowed more expensive options occasionally to be accepted. The production volume was set to 200 000 units, to allow comparison with the global minimum calculated from the exhaustive solution set.

Table 4.13 Test strategy selected by simulated annealing

Functional block	Testability method	Accessibility method
Node 1 (PLA)	treuer	ext_scan
Node 2 (Random sequential)	int_scan	ext_d_latch
Node 3 (Random sequential)	int_nodft	ext_nodft
Node 4 (RAM)	illman	ext_nodft
Cost : 5 706 739 DM		

The resulting strategy, shown in table 4.13, is approximately 0.3% more expensive than the global minimum. In fact, the strategies are very similar, but an additional accessibility enhancing method is selected for the PLA which in accessibility terms is redundant. This kind of redundancy is very simple to spot and remove as a post-processing step.

The progression of the simulated annealing test strategy selection is shown in figure 4.16. A total of 80 strategies were attempted. The graph shows the number of attempts in each cycle before the control parameter T is reset. Crosses represent accepted solutions. The large cost values occur when an exhaustive test vector set is chosen for the PLA.

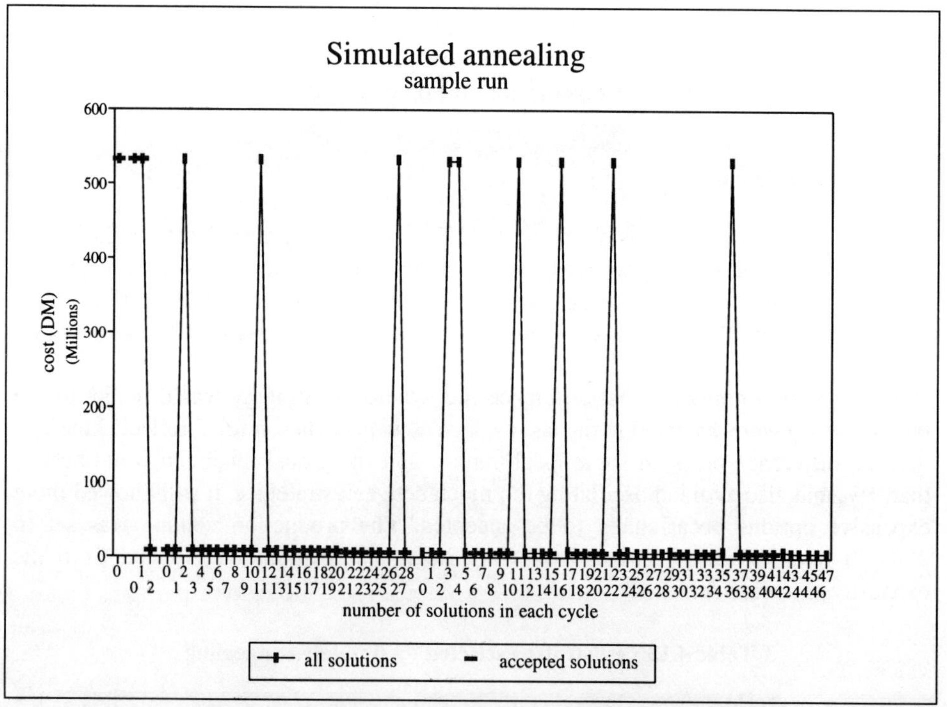

Figure 4.16 Test strategy selection by simulated annealing

Figure 4.17 shows the lower portion of the graph. In the final cycle, all the accepted solutions are in the lowest price band of the test application costs. In fact, the steps in overall cost that are due to test vector application prices are visible in the way groups of solutions are accepted.

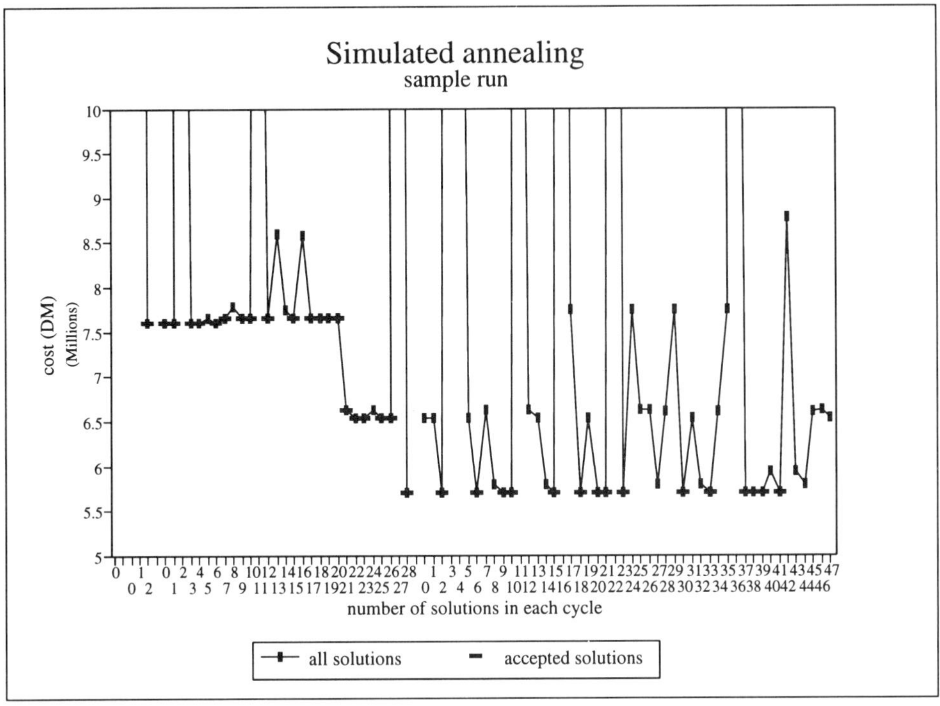

Figure 4.17 Test strategy selection by simulated annealing (detail)

4.8 ATE AVAILABILITY EVALUATION

In addition to the test strategy planning described above, ECOtest can be used to evaluate a range of alternative scenarios which are not modelled explicitly. One example is the availability of scan ATE. This is assumed to be the case in the current model. However, it is clear that ATE availability may influence the choice of test strategy. Indeed for large production runs a new ATE could be considered for purchase if it reduced the overall cost. The following example shows the results from a DFT selection procedure for the example circuit (demo_circ) if a scan ATE is available. It should be remembered that different scan ATE will have different pricing functions.

For 5 000 ICs the total design manufacture and test cost was 522 570 DM, and the test strategy selected was as follows:

Node	**Node type**	**Test method**
Node 1	PLA	hfc
Node 2	Random seq	int_scan
Node 3	Random seq	int_scan, ext-scan_set
Node 4	RAM	illman

For 500 000 ICs the total design, manufacture and test cost was 13 565 278 DM, and the test strategy selected was as follows:

Node	Node type	Test method
Node 1	PLA	hfc
Node 2	Random seq	int_scan
Node 3	Random seq	int_nodft, ext_d_latch
Node 4	RAM	illman

However, if a scan ATE is not available and the same test strategies are applied using a non-scan tester then a penalty is incurred. For 5 000 ICs the new total cost is 547 570 DM, an increase of 25 000 DM (5 DM per IC). However, if the test strategy planning procedure is performed again with the knowledge that no scan ATE is available, in this case no 'better' test method can be selected. Thus the DFT method selected does not change but there is an extra cost incurred due to the lack of a scan ATE. The single cost incurred from this production run is not enough to consider purchasing a scan ATE. If all the production runs of a company over several years are considered the position might be different.

In the case of the 500 000 production run, however, the increase in cost is significant. If a scan ATE is not available and the same DFT strategy is used, then the total cost is 16 054 360 DM, an increase of 2 489 082 (4.88 DM per IC). If we consider the fact that no ATE is available when the DFT selection process is performed then a new test strategy plan is obtained. The cheapest test strategy for 500 000 ICs when no scan ATE is available is shown below:

Node	Node type	Test method
Node 1	PLA	treuer
Node 2	Random seq	int_nodft
Node 3	Random seq	int_nodft, ext-d-latch
Node 4	RAM	illman

The total cost is now 15 805 150 DM, an increase of 2 239 872 DM (4.48 DM per IC). There is still a large penalty associated with not having a scan path tester but it has been minimised by reconsidering the test strategy plan. It can be observed that the new test strategy relies less on the use of scan paths and more on self-test and traditional non-intrusive test methods.

5

Test strategy planning and test economics for boards

5.1 THE IMPACT OF TEST STRATEGIES ON THE ECONOMICS OF ELECTRONIC SYSTEMS

So far, we have concentrated on how test decisions affect the economics at chip level. However, VLSI test strategies also affect the economics of higher levels of assembly, and later stages of the product's life cycle, and can be used at subsequent test levels. For example, a self-test which is designed for component testing can also be used for field test and diagnosis. If for such a self-test only the costs at component level are evaluated, the result might be that the self-test is less cost effective than other test strategies. If *all* economic aspects during the product life cycle were considered, the conclusion might well be reversed. Thus an economic evaluation of test strategies should be done for all test-sensitive cost areas of the product's life cycle. In terms of the economics model, the scope of *test dependent parameters* should be extended to cover all stages of a product's life cycle.

Similarly, the economic evaluation of test strategies such as boundary scan, for a system or for a board, should take into account all costs which can be affected by the test strategy. This section will identify the effect of test strategies on the life cycle of electronic systems, and develop a method to link all test dependent parameters to the economics model.

Most test methods influence almost all phases of the life cycle. Consider an example: it is decided that *boundary scan* will be used on the board for a new electronics system. The use of boundary scan will influence the design cost of the components, the boards and the entire system. It will also affect the production cost of the components, the board and the system. It will have an impact on the test and diagnosis of the board and the system (but typically not the component). And boundary scan can be used for field diagnosis, and will therefore affect the down times of a system in the field and the field diagnosis cost. In addition, boundary scan can increase the product's quality, which means that the mean time between failure decreases, and again the down times are decreased. Also, time to market may be shortened by implementing boundary scan. This is because of a lower risk of a redesign due to lack of testability of the design, and because of lower fault diagnosis times during development, which can easily exceed one week for a complex VLSI based system. All these phases must be considered in evaluating the economics of boundary scan.

The advantage of a life cycle cost analysis is that a test strategy is not limited to the combination of test methods for a part of the design, or a particular life cycle phase of the system. A *global test strategy* can be optimised for the whole system in all phases of the life cycle. This leads to much better results from an economic point of view than optimising a test strategy just for a part of the design of life cycle. But the main objective of a test strategy is to get a testable, working system rather than to achieve a specific fault coverage for only a part of that system. Additionally, it would be expected that the results of the analysis would be better accepted by the people who are expected to implement it.

5.2 STRUCTURE OF A TEST ECONOMICS MODEL

In order to quantify the impact of test (and design) decisions on the economics of electronics systems, some detailed cost modelling is needed. It must be stressed that this type of cost modelling is company dependent. It is naïve to expect that an 'off-the-shelf' cost model (if one can be found) can be used without modification to make financial decisions. The rest of this chapter will provide some pointers towards economic model design, together with a system developed by the authors as an example of an economics based software tool to aid the test strategy planning process.

The impact of a test method on the economics of the system's life cycle is described by the *test dependent primary parameters*. The values for these parameters are different from test method to test method and therefore they are different for alternative test strategies. Besides the test dependent primary parameters, the test economics model comprises *design dependent primary parameters* and *design independent primary parameters*. Figure 5.1 shows the elements of the test economics model. The secondary parameters are based on the primary parameters. Primary parameters are provided from different sources:

- The *design independent primary parameters* are those which do not vary from design to design. Their values depend on the design environment (such as the productivity of the CAD system used), or they are simply normalising factors. The values of these parameters are stored in a data base which can be updated whenever the values change. The values are company specific and once they are set, they will rarely change.
- The *design dependent primary parameters* are those which vary from design to design and which are test independent (e.g. the production volume). Their values are provided through a CAD interface.
- The *test dependent primary parameters* are provided by test method dependent cost models. This means that for each test method a separate cost model is developed, which calculates the test dependent primary parameters from design dependent primary parameters and test method dependent primary parameters. This means, that the structure of the entire economics model depends on the test strategy, because parts of the economics model depend on the test methods which are applied.

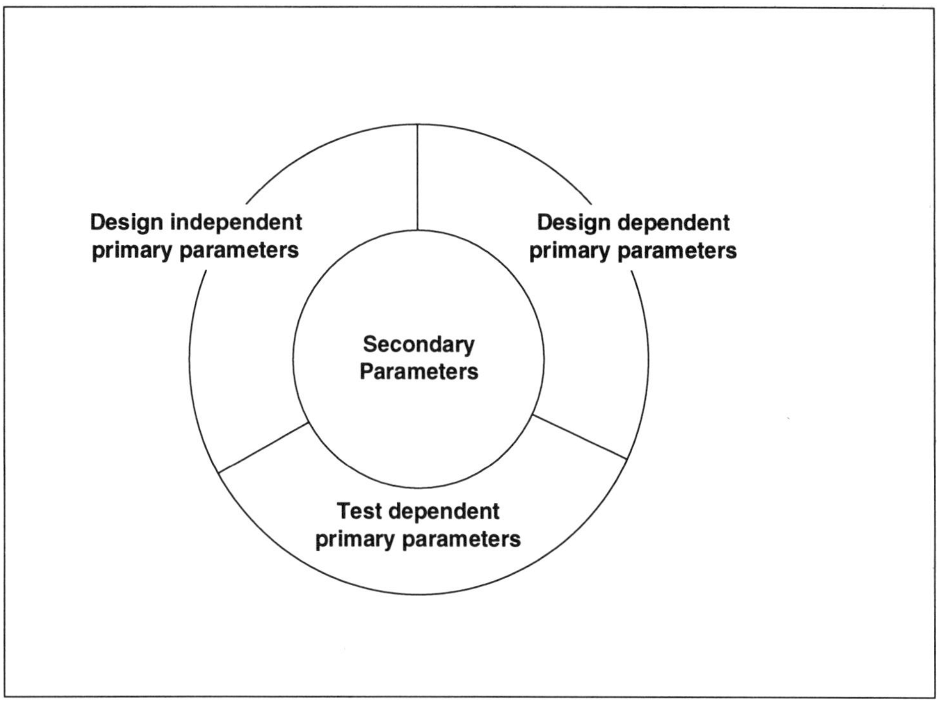

Figure 5.1 Classification of test economics model parameters

The test economics model can be partitioned into several cost sections. The partitioning process will be driven by two criteria:

- The life cycle of the system will be partitioned into several phases and sub-phases. The phases will be derived especially from a test view.
- The electronic system will be partitioned from the assembly and production view. This means that every level of assembly will be represented by a separate cost section with its own life cycle phases.

So the test economics model can be seen as a two-dimensional model. One dimension is the time (the life of the system), and the other dimension is represented by the parts of the system. In order to set the modelling in context, the life cycle phases and levels of assembly of the system need to be defined. Again, these are not fixed, and the details vary between companies. The definitions here simply establish a baseline which the authors used for their economics models.

The electronic system is built upon four levels of assembly:

- The whole *system* includes all parts needed to perform the specified functions.
- The system is composed of several *boards*. A board is defined as a module containing several electronic components or sub-modules, which can be assembled into one unit. The connection of boards within a system is implemented by plug-in connections.

- A *module* is composed of electronic components. A module is simply a sub-assembly of a board, and for the test economics model, it will not be seen as a separate assembly level.
- A *component* is the smallest unit for assembly. Electronic components are grouped into passive/active and digital/analog/hybrid. Digital components are usually produced as ICs. In VLSI based systems, the essential components are VLSIs. They may be designed in-house, and different test strategies may be evaluated for them. Therefore a separate test economics model is developed for VLSIs. The other components are assumed to be purchased, and only the purchase costs per component are considered.

The test economics model is partitioned into the system level, the board level and the component level, where the costs of each level are included in the model above it. Table 5.1 outlines the phases of the life cycle of electronic systems.

Table 5.1 Life cycle model for electronic systems with regard to test

Phases	Sub phases	Areas of cost origin
Formation	Initiative	
	Planning	Test strategy planning and decision making
	Development	Design, prototype production, verification test engineering
Implementation	Production	Purchase, manufacture and assembly
	Test	Test, Diagnosis, Repair
Usage		Installation, Maintenance, Field diagnosis and repair, Logistics
Phase-out		

▩ phases are not considered

The partitioning into sub-phases was performed with regard to the impact of test strategies. The test economics model can be simplified by neglecting phases not relevant to the economic analysis. These are the initiative phase and the planning phase, because the related tasks are performed *before* the test economics model is used, and therefore related costs are not influenced by the test strategy. The phase-out phase does not include test strategy dependent cost (see previous section), and therefore the related cost can also be neglected for the test economics evaluation.

All other phases are considered for the test economics model. Some of the phases are included at all levels of assembly, some are included only for a subset of the assembly levels.

5.3 A LIFE CYCLE TEST ECONOMICS MODEL

The test economics model for VLSI devices and VLSI based systems and boards is based on the structure described in the previous section. Each level of assembly consists of a set of cost models, which are related to the life cycle phases of the assembly unit. Costs occurring in the field have been separated from the other phases, as they relate to the entire system. The total cost per assembly unit is derived summing the cost of each phase, and the total life cycle cost is derived by including the cost of lower levels of assembly. The economic data needed in the model from a lower level of assembly are grouped into three classes:

- The *volume related costs (VRC)* are expenditures which occur per device produced.
- The *non-recurring costs (NRC)* are expenditures which occur once per product type.
- The *investments* are expenditures which can be shared with other products.

5.3.1 The test economics model for boards

The test economics model for boards includes the two main phases of development and production. Figure 5.2 shows a flow diagram of the phases.
The development phase includes the following sub-phases:

- The design phase comprises the initial design, design entry and computer simulation.
- The layout phase includes the placement and floor planning of the printed circuit boards.
- The prototype manufacture covers the construction and manufacture of prototype boards.
- The verification phase covers the evaluation of the board by verifying the specified functions with the prototypes.
- The test engineering phase covers the generation of test patterns, the generation of the test programs and the manufacture of test tools, such as a bed-of-nails fixture for an in-circuit test.

The production phase is partitioned into two sub-phases in order to simplify the test-related modelling:

- The manufacture phase includes the production preparation, fabrication and assembly.

- The test phase comprises test application, which includes the costs for test, diagnosis and repair.

In the test economics model structure the test engineering phase and the test phase are combined into a single cost model because the cost structure of these costs is test method dependent. For each test method a separate cost model exists, which models the related costs for test engineering and test application.

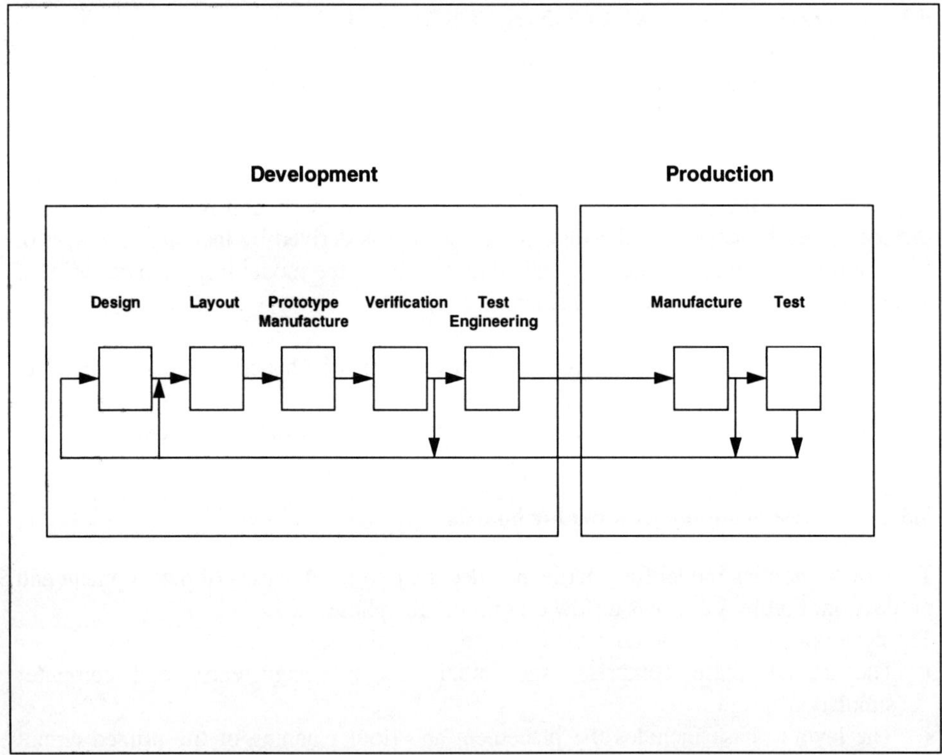

Figure 5.2 Flow diagram of board level phases

The life cycle of a board includes a chance that a redesign may become necessary. This can happen as a result of the verification or during the production phase. This fact is considered in the test economics model by the definition of the probability of a redesign, and by iteration factors per development phase, which define the effort needed for the redesign as a percentage of the original effort.

The costs in the development phases include labour costs, equipment costs and material costs. The labour costs are determined by the hourly labour rates and the predicted effort. The equipment costs are related to the computer equipment, which is needed for the development of the design. They are also calculated from the hourly costing rates and the estimated time the equipment is needed. If the costs are included in the hourly labour rates of the designers, the equipment costs can be set to zero. Material costs occur only in the manufacturing of the prototypes.

The manufacture costs are composed of the production preparation costs, the material costs and the assembly costs of the board. The material costs include also the total costs for the components, which are calculated by the ASIC test economics model. The calculation of the assembly costs is based upon the assembly cost per component and per assembly type and the number of components per assembly type. In addition to the costs the number of defects per defect type is calculated. These data are needed for the test phase.

The test cost model includes the calculation of the test engineering costs and the test application costs. The test engineering costs are composed of test tool manufacture costs and test generation costs. The calculation of the test generation costs is based on the engineering effort, the engineering labour rate, the usage of equipment and the equipment rate. The test application phase is built upon the test phase and the diagnosis phase as shown in figure 5.3.

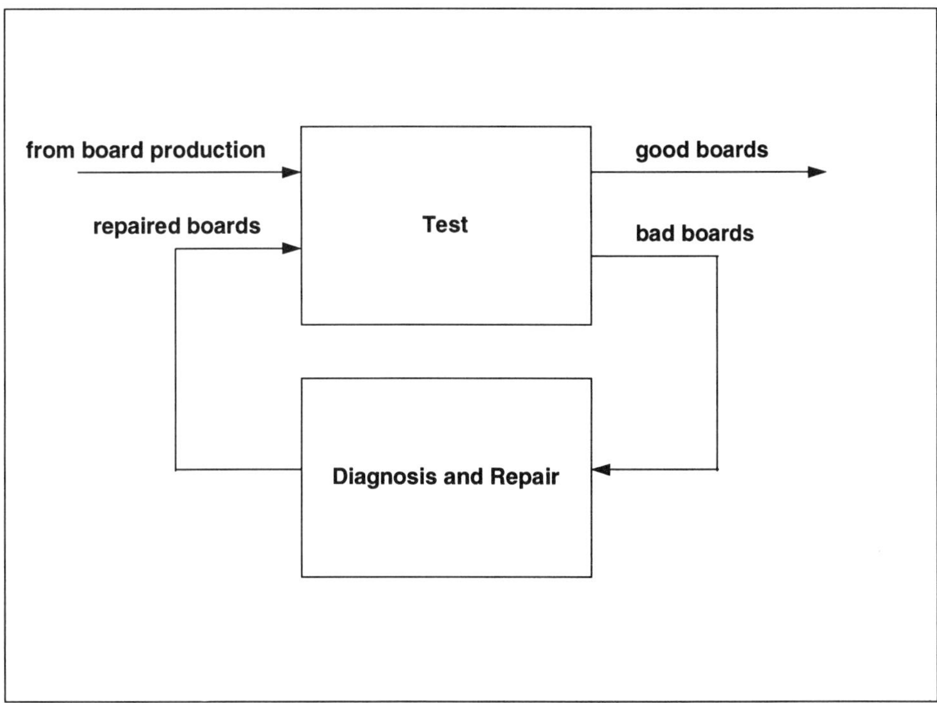

Figure 5.3 The test/repair loop of the test application phase

The test is applied to all boards coming from production. The diagnosis and repair are applied only for those boards in which a defect has been identified during the test. All repaired boards are re-tested in order to detect multiple defects and to detect defects which arise during repair.

In addition to the test and diagnosis/repair partitioning, the test economics model is partitioned into the following parts:

- In the quality model the fault coverage of the test is calculated, which is based on the fault spectrum of the previous stage. Depending on the test strategy, the previous stage can be the manufacture phase or another test phase.
- The time model calculates the total times for test application, diagnosis and repair, depending on the times per board and the number of boards going to test and to diagnosis/repair.
- The cost model calculates the actual financial cost. This includes the procurement and the usage of test equipment, and the labour costs for the test personnel. In addition, the repair costs are calculated. These are based on the labour costs and the material costs. In the calculation of the material costs the type of defect is taken into account.

The times and the financial costs are calculated for the test phase and the diagnosis and repair phase.

5.3.2 The test economics model for systems

The test economics model for electronic systems includes the phases of development, production and field usage. The field usage will be described in the next sub-section. This section outlines the models of the development phase and the production phase of the system.

In the area of *system development* only the test engineering costs are relevant for the test strategies. The test strategy dependent development costs are mainly related to board development and component development, and the related costs are included in the total system costs as part of the total board costs. The test engineering costs are combined with the test application costs in the same way as for boards.

The *production cost model of the system* includes the costs for the assembly of the boards into a system, the total costs of the boards and optional device costs for incorporated test devices, which support the test and diagnosis of the system in the field.

The system test costs comprises test engineering costs and test application costs in the same way as for board. The test/repair loop is the same as for boards.

5.3.3 The test economics model for the field costs

The test relevant parts of the field phase of the life cycle for electronics systems are the installation of the system, the field maintenance and the field breakdown. All three phases consist of a diagnosis/repair loop. The main difference between the phases is the test phase. The installation test is different from the maintenance test, and the breakdown phase does not include a test phase, because the system is known to be defective. But in the case of a breakdown, the system is always defective and goes through the diagnosis and repair loop, whereas in the installation phase and the maintenance phase only a portion of the systems go to diagnosis and repair.

The diagnosis and repair phase consists of three stages. Figure 5.4 shows the field repair loop. If a system is defective, the defect is diagnosed in the field. If the defect can be detected and repaired in the field, it will be repaired there. If not, the system will be repaired in the field by replacing the defective boards, and the defective boards go to the service centre for further diagnosis. The service centre provides the spare boards. If the defective board can be repaired in the service centre, it will become a spare board. If not, the defective board will go into the depot, which is mostly identical to the production facility. If the board can be repaired in the depot, it will be sent back to the service centre to go into the spares stock. If it cannot be repaired, it will be discarded.

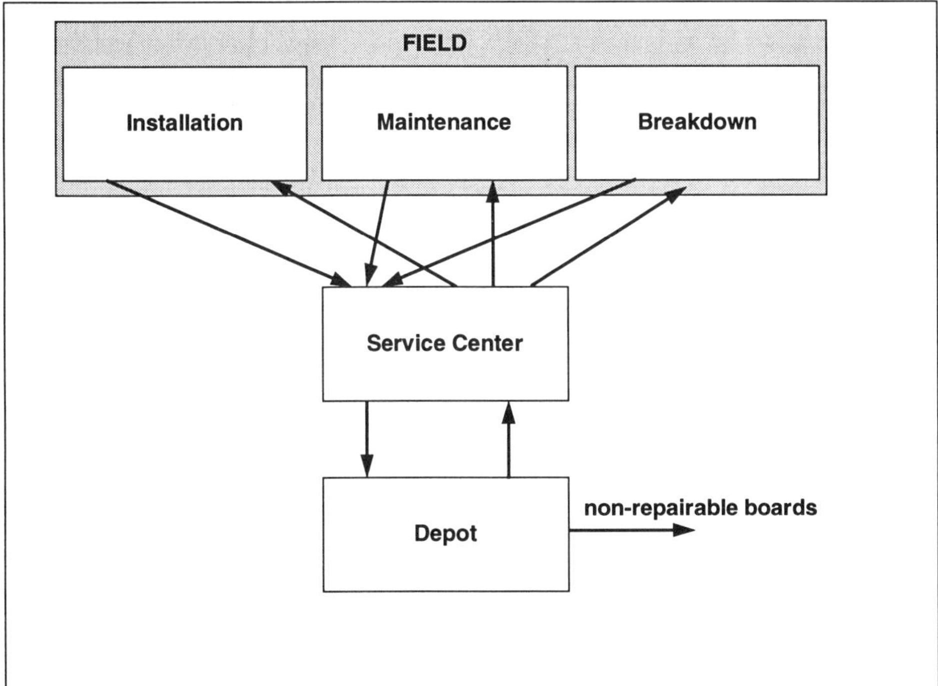

Figure 5.4 Field repair loop

From this scenario the following cost models are defined:

- The *field usage cost model* consists of field installation, field maintenance and the field repair costs. This includes the defect rates of the system to be installed, the defect rates of maintained systems and the mean time between failures in order to calculate the breakdown rate. The number of systems which cannot be repaired in the field is determined from the field repair rate, and the number of boards going to the service centre is derived from the average number of boards which are replaced at one failure and the number of failures.

- The *service centre cost model* includes the costs for the provision of spare boards and the repair costs for boards which are actually repaired in the service centre. The number of spare boards which are needed takes into account the time needed to repair a defective board.
- The *depot cost model* includes the repair costs for the boards and the production of new spare boards to replace non-repairable boards. The number of non-repairable boards is derived from the total number of boards going to depot for repair and the percentage of boards which are not repairable.

5.3.4 Consideration of interest rates

Interest rates can be an important factor if the time for return-of-investment is very long and if market interest rates are significant. In these circumstances the early development phase costs will differ from the same costs during production. Consider an example.

The total development time of a system is two years. The market interest rate is constant at 12%. The initial design is performed at the beginning of the development phase, and test program generation is performed at the end of this phase. For simplicity's sake, these two phases are assumed to be very short compared to the total development time. The costs for each phase are £10 000. To compare these two costs, they must be related to the same time point. We take the end of the development time as the common time point. This means that, taking the interest rate into account, the £10 000 for the test engineering phase remains unchanged, whereas the £10 000 for the initial design becomes £12 544.

So, expenditures which are made early are more expensive than the same expenditures made later. This fact can be important for test strategy planning, if the interest rates are high and two test strategies require expenditures which have to be made at significantly different times.

For that reason the interest rate aspect was included in the model. All costs are converted to a common time point, which is the end of the life cycle, using the present value method [Blohm88]. This approach can be varied, and different return-on-investment calculations can be employed. However, the above approach was found to be adequate for our purposes.

5.4 USING ECONOMICS MODELS TO PLAN BOARD TEST STRATEGIES

The authors developed a test strategy planning tool using the economics models outlined above. The tool (called ECOvbs for *eco*nomics of *V*LSI *b*ased *s*ystems) was developed under the European Community ESPRIT programme, and was the result of a collaboration between Siemens-Nixdorf Informationssysteme AG (SNI) and Brunel University. This was a continuation of the effort which resulted in the test strategy planning tool for VLSI chips, ECOtest. The tools was intended for industrial use, and the following requirements were considered:

- The system should take into account industrial practices and needs.
- The system should profit from the experience gained in manual test strategy planning by quality assurance people.
- The system should be integrated into an industrial environment.
- This was a novel approach, and the authors were aware that the system should be a good compromise between the state of the art in research and the applicability of the system in industry.

In the rest of this chapter the philosophy of the test strategy planner and the emerging need for such a system in industry will be discussed first. In the subsequent sections, an overview of the architecture and an implementation outline of ECOvbs will be given.

5.5 PHILOSOPHY OF ECOvbs

In the early 1980s manufacturing efficiency became the cornerstone in industry with the main objective being *the highest quality product at the lowest possible cost* (see [Pynn86]). Pynn states that 'today's successful business requires an objective which takes both product quality and manufacture efficiency into account' [Pynn86], and based on this statement, he defines the following manufacturing objective: *To develop a production strategy which will achieve a product with superior quality manufactured at the lowest possible cost.*

One crucial part of the production strategy, especially in the field of VLSI based systems, is the test strategy. The test strategy is the essential factor for quality. And it affects nearly all cost areas of an electronic product.

Pynn [Pynn86] defines a test strategy as follows: *A successful test strategy is the optimum arrangement of various testers in the circuit board manufacturing process that will result in products of maximal quality at minimum cost.*

Based on this definition, the technical factors affecting the test strategy are the fault spectrum, the process yield, the production rate or volume and the product mix, i.e. aggregation of active board types in a production line. The economic factors are the tester acquisition cost, the tester adaptation cost (fixture cost and test generation cost) and tester operation cost.

In this approach it is assumed that for a given test strategy the technical factors are fixed, and a test strategy is built purely upon a mixture of test applications. A similar definition of test strategies is made in [Davis82]. This definition of a test strategy makes the test strategy planning task relatively simple for the following reasons:

- The only degree of freedom in the strategy planning is what test equipment should be used in which phase of the production cycle.
- Therefore the cost areas affected by a test strategy are only the cost related to the test equipment, as defined in the previous paragraph.
- Decisions about the test strategy can be made rather late. The first phase in the product life which is affected by the test strategy, is the test generation phase, which in the past was done after the design cycle.
- There was no great choice of test equipment. Automatic test equipment use started at the end of the 1960s, and only a few companies shared the market in test equipment.
- The rate of change in technology and therefore test technology was much lower than today.

These factors were the main reasons why no need was perceived for a test strategy planning tool. The task of test strategy planning was confined to the selection of a new test equipment or establishing whether the existing tester scene was appropriate for the new product.

Today the situation has completely changed:

- It is now accepted that quality and production costs are mainly determined during the specification and design phase of the product. [Reinertsen83] claims that 'decisions on technology, architecture and physical structure, which typically occur during the first few months of the design process, determine 90% of the product cost and the majority of its capabilities'. A similar statement is made by [Szygenda92]: 'although the system design phase of product development represents only about 15% of the product's cost, it has a 70% impact on that product's operation and support cost'
- Today, design methods are available to control production and quality. Design for testability (DFT), design for manufacture (DFM) or concurrent engineering are the key phrases in this field.
- Because of rapid technology changes, the product lifetime becomes shorter, and this fact and an increasing competition in the market lead to enormous price pressures in the electronics market. This brings back the need for low production costs right from the beginning of the production.

For these reasons, today's test strategy includes, besides the usage of test equipment, the DFT and DFM methods. It is clear that an alternative definition is needed.

A test procedure consists of the provision of the test equipment and the environment to utilise it, the adaptation of the test equipment to the device under test, i.e. the hardware adaptation and the test program generation, and the test application. The design methods are those which facilitate the realisation of the test procedure (DFT) and those which support the avoidance of errors, or, more specific, which increase the yield in the production (DFM).

The optimum test strategy can then be defined as follows: *An optimum test strategy is a test strategy which meets all given product constraints and the system's final quality requirements by minimum total cost.*

Based on this definition and the changing technology, design, production and market in electronic systems, the determination of the optimum test strategy and therefore the test strategy planning task is much more complicated than an approach based on a test strategy as defined by Pynn [Pynn86]:

- The test is now completely integrated into the design process. Therefore the decision on the test strategy must be made very early in the specification phase of the product. This makes the derivation of the test strategy factors, such as the prediction of yield, fault coverage, production cost or test adaptation costs much more complicated and uncertain. Especially in an environment with quickly changing technologies, it becomes very difficult to estimate for these factors, as we cannot depend on previous experience.
- The estimation of the total cost is much more complicated than the estimation of the economic factors as defined by Pynn, which are purely related to the test procedure. Today's test strategies affect cost areas such as design cost for different design alternatives, component procurement, cost of the production process, test generation cost or test, diagnosis and repair cost in production or in the field. All these costs have to be determined in a product phase, where factors such as the production yield, production volume, design complexity, fault coverage of a test or the fault spectrum cannot be measured and therefore are based on estimates.
- The multitude of test strategies is increased by another dimension, which is the design option.
- Factors such as the fault coverage per test and the production yield are continuous parameters of the test strategy, which need to be optimised per test strategy to be analysed. This optimisation process (which is the optimum mix of yield and fault coverage in order to achieve a given product quality?) can be in itself an extremely complicated task.
- The planning of test strategies will need organisational changes and changes in responsibilities in most of the companies, because the implications of a test strategy are no longer limited to the production people (see also [Davis92]). The implementation of a test strategy as defined here involves most parts of the company: engineering, manufacture, finance, service, purchase and even marketing and sales (e.g. to determine what quality level the market requires).

All these aspects necessitate structured support for the test strategy planners. This support can be given by a software tool, which supports the user in

- providing all factors which are needed to determine the optimum test strategy
- evaluating the test strategies concerning its economics and compliance to quality and design constraints
- optimising the parameters of a test strategy
- selecting the optimum test strategy.

The development of ECOvbs attempted to address the above points. ECOvbs was implemented in C++.

A test strategy for VLSI based systems includes test procedures at all levels of integration. Therefore the test of the components and the related test strategy is part of the test strategy of the entire system. Also, for several DFT methods which are implemented at component level, or which are used for component level test, gains are achieved for board level or system level test. In particular, boundary scan is implemented in the component for supporting the board level test. Other examples are the scan path technique, which was originally used and demanded by the system test people for diagnostic purposes ([Sedmak92]), or built-in self-test techniques, which are used for board test and diagnostics. These aspects make the test strategy planning task hierarchical, and the test strategy planning system for VLSIs, ECOtest, can be integrated into ECOvbs for the test strategy planning task at component level for the VLSIs.

5.6 SYSTEM OVERVIEW

ECOvbs provided the user with an estimate of the costs involved in using a particular test strategy on a given electronic design. The user could look at cost-quality trade-offs and could acquire an understanding of the fault spectrum of the electronic system at each stage of testing.

In order to provide all the support needed to make test strategy decisions, ECOvbs comprised the following features:

- A series of cost models are used which describe the cost structure for all cost areas which are affected by test strategies.
- Cost parameters are linked across the cost models.
- The cost calculations are fully parametrised instead of using rules of thumb.
- ECOvbs utilises a user supplied design description of a board.
- The test method descriptions and the test equipment descriptions are provided as text files so that they are available for general usage.
- The test strategies can be set up by the user to consist of one or more test stages, which can test different parts of the electronic system, or which are specialised to detect different fault types.
- Test clusters can be defined in order to apply specific test stages only to parts of the electronic system.
- The defined test strategies can be stored for later reference.
- ECOvbs automatically generates the test strategy specific cost models and creates the linkage between the cost models.
- An integrated verification function checks the correctness and completeness of test method descriptions, the applicability of the test methods to the design, and the consistency of the test strategy.
- Test strategies can be varied by deactivating particular parts of the test strategy.
- Based upon the defined test strategies, the ECOvbs system enables test strategies to be evaluated in parallel in order to facilitate direct comparison of the cost components.

- There is considerable flexibility in looking at the costs at each test stage.
- ECOvbs calculates the fault spectrum after each test stage.
- ECOvbs provides a function which graphically shows the fault spectrum and the total costs per test stage.
- The results can be stored in files for printing and for later reference.
- The user interface provides general commands to execute macros or to administer the data in addition to the commands to execute the functions which are listed above.
- Extensive help support is provided in order to facilitate the usage of the system.

Figure 5.5 gives an outline of the architecture of ECOvbs.

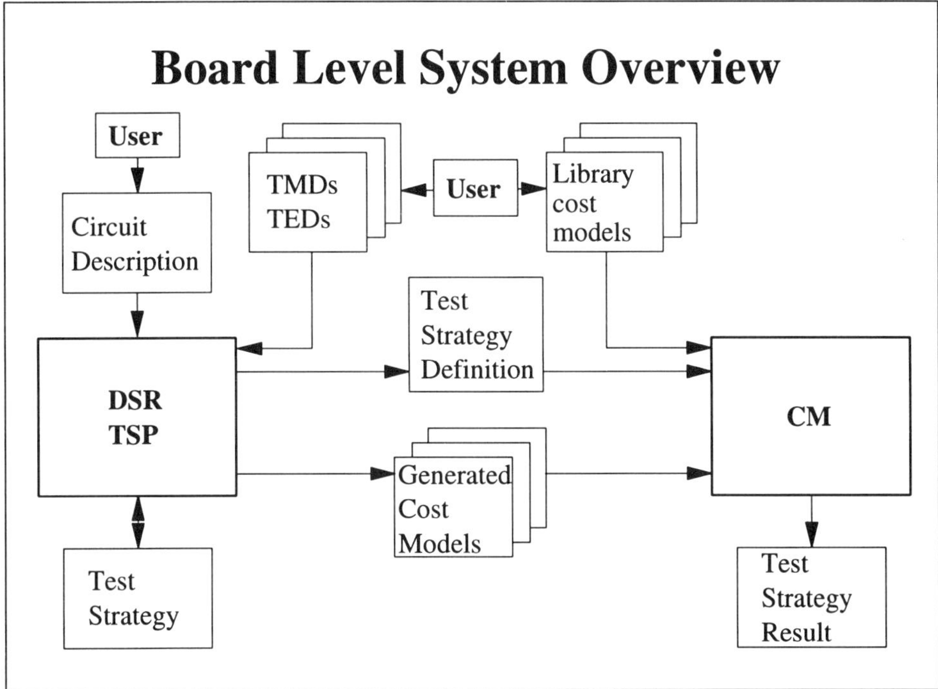

Figure 5.5 Architecture of ECOvbs

The design specification reader (DSR) handles the design data. It allows the design specification to be read and displayed. The read function checks the design description for correctness.

The test strategy planner (TSP) is used for managing the test strategies. Based on the design description and the test method / test equipment descriptions, it provides the user with the following functions:

- Read the test method and test equipment descriptions.
- Read and write the test strategy creation file.

- Create test strategies.
- Generate the test strategy specific and design specific cost models and the test strategy definition file.
- Verify the test methods and test strategies.
- Activate and deactivate test stages of a given test strategy.
- Print test method descriptions, test equipment descriptions and test strategies.
- Administer the test strategy related data, which are the test strategy creation file, the test strategy specific cost models and the test strategy definition file.

The cost model handler (CM) is used to evaluate the cost and quality data for a given test strategy. For that purpose it provides the following commands:

- Load and delete the cost models and the test strategy definition file for a given test strategy.
- Print cost model data for the test strategies loaded.
- Draw the cost model structure for a test strategy.
- Draw the sensitivity of one costing parameter to another costing parameter.
- Draw the fault spectrum and the costs per test stage.
- Set a cost model parameter to a certain value and evaluate this impact.
- Reset all cost parameters to its initial values.

When quitting ECOvbs, all the information about the test strategy (design requirements, test stages, main costing data) is stored in a result file.

5.7 THE ECONOMICS MODEL IMPLEMENTATION

In implementing the economics modelling for ECOvbs, flexibility was a major consideration, as a variety of test strategies (comprising different test stages) had to be evaluated. The economics model is in fact a collection of many basic models that can be combined in different ways to create a single economics model for a specific test strategy. This has two main advantages:

- This structure facilitates easy modifications and verification of the model. For example, if the production environment changes, only a few small models need to be changed. In addition it makes data acquisition and model generation a more streamlined procedure, as individual experts can address each area of the model.
- Many of the basic models can be reused several times to build up a complex model. For example, models that define a test stage can be repeated as more test stages are added, with only minor changes made to the input data. This also has the advantage of minimising the errors in the modelling process. If the model is intended to be used to compare the relative cost of test strategies and not to calculate accurately the absolute design, production and test cost, then it is more likely that any error in absolute cost will not affect the relative cost of test strategies.

To indicate how the collection of basic cost models interact a simplified structure is described below.

Board data Physical design parameters and components on the board

Design data	Design environment models and human effort data
Labour rates	Basic labour models used to calculate labour costs
Design costs	Addition of basic models and other parameters to calculate design costs
Iteration factors	Prediction models to calculate the iteration factors for the design layout and verification loop
Production data	Models to calculate the production environment and effort
Production cost	Addition of basic models and other parameters to calculate production cost
Assembly model	Contains models that detail the cost of assembling different types of components
Fault spectrum	Models to calculate the number of faults per board of each fault model type
Defect-to-fault	Models to calculate the likely faults from different manufacturing defects
Test stage/ATE	Test generation and throughput models for a test stage.

Figure 5.6 shows the basic model when there are no test stages. Figure 5.7 shows the basic model interaction when a single test stage is added.

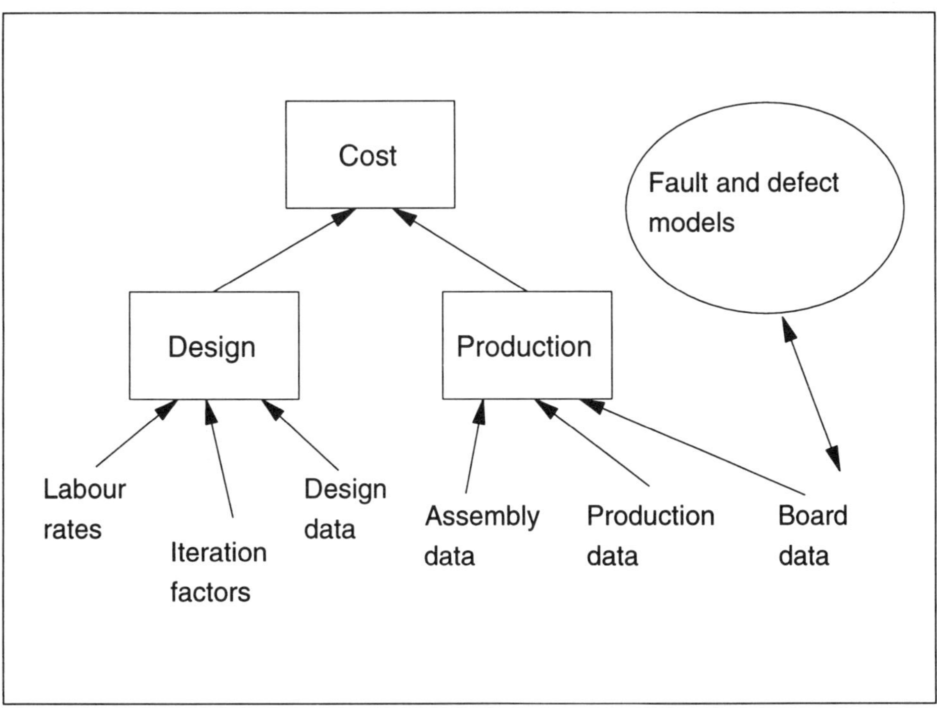

Figure 5.6 Cost model with no test stage

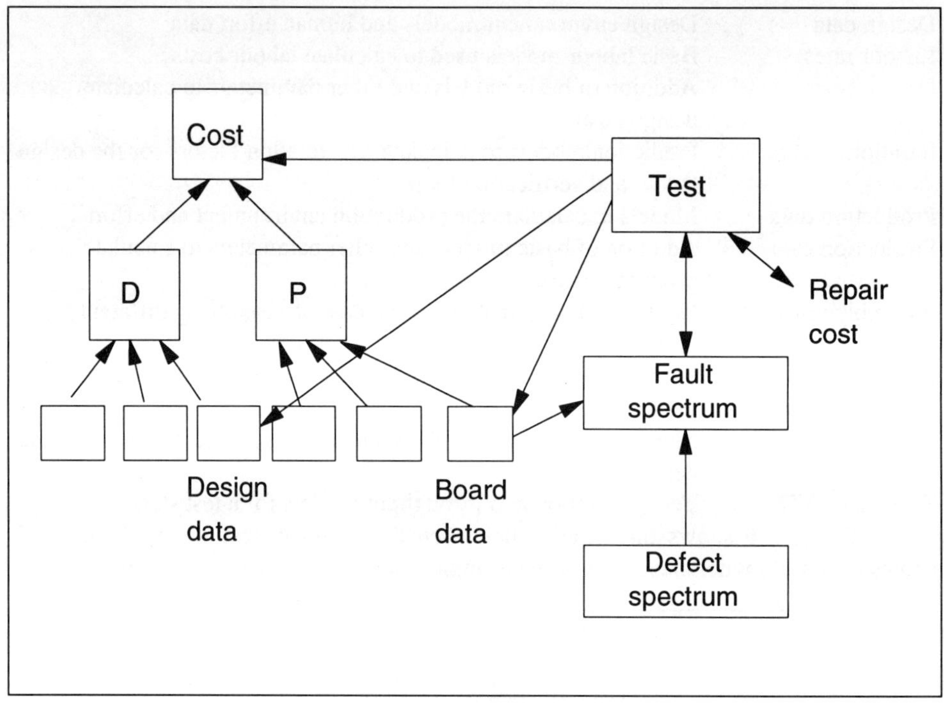

Figure 5.7 Cost model interaction when a test stage is added

It can be observed that as a test stage is added the following interactions are possible. Firstly, there are the test generation, preparation and throughput costs to be added to the overall cost. In addiction, any modification to the board design could affect the design data. For example, the test strategy might require test sites or different components (e.g. boundary scan) to be added, and this will affect the design and layout of the board. This extra engineering time and effort will affect the design stage cost and the overall cost. In addition, the use of different or additional components might affect the assembly cost. The number of board connections is also likely to change. The addition of different components and an increased number of connections will modify the fault spectrum as the likelihood of failures is modified. Thus the testing/repair throughput cost is modified. The repair cost might need to be modified to take into account the time and effort for different failures and cost of replacement items. As more and more test strategies are added so the interaction and nesting of the cost model are increased.

5.8 CALCULATION OF FAULT SPECTRUM AND DEFECT SPECTRUM

The above should provide a good picture of the capabilities and implementation strategy of ECOvbs. However, apart from costs, ECOvbs provides detailed quality calculations by propagating defect and fault spectra through test stages. The calculation of these spectra is worth a detailed look, and is described in this section.

The terms 'defect' and 'fault' are often used for the same effects. However, there is a difference in the meaning of the terms. Therefore we will introduce the definitions of these terms, as they are used. Similar definitions are given in [Bennetts89]: *A **defect** is a discernible physical flaw which causes the device not to work correctly under certain conditions. A **fault** is the effect which is caused by a defect, and which can be measured by a test system.*

As an example, a broken wire on a bare board is a defect, which will lead to an open fault.

Defects are related to the manufacture and repair phase, because defects come into existence during the manufacturing process, and are eliminated by means of a repair. But in most cases - apart from gross defects - a defect cannot be identified directly. For these cases, test systems are used to measure the fault, which is related to the defect. If the defect should be repaired, a fault diagnosis is performed in order to isolate the fault, and to relate it to the actual defect.

Due to the complexity of today's electronic systems, and therefore the complexity of defects and faults, there are special test systems which are optimised for the detection of certain fault classes. In order to predict the fault coverage of a test method, it is important to know the fault rate per fault type of the device under test. These data are known as the *fault spectrum.* The fault types which form a fault spectrum depend on the product mix. A typical fault spectrum consists of opens, shorts, static faults, dynamic faults, voltage faults and temperature faults.

Now the question remains how to predict the fault spectrum of a system which has not yet been designed. In previous work this prediction was always made by taking the numbers from the systems which are already in fabrication, and for which the fault spectrum can be measured. But this method implies two problems which can make the prediction unreliable:

- The faults which are measured in a production process are not necessarily the faults which the device under test contains, because only those faults are measured which can be detected during the test. But the detection of a fault depends on the fault coverage of the test. Hence the measured fault spectrum depends on the fault coverage of the tests. In many cases the assumption is that the total number of faults in a system is the total number of faults detected at all the test stages during production.
- If a system is based on a new technology, it definitely cannot be assumed, that the fault spectrum will be the same as for the system, which is based on an older technology. This is because the production process, the component defects and the composition of the system will be completely different.

For these reasons, a prediction of the fault spectrum by measuring the fault spectrum of the previous system can be very inaccurate, especially if a new technology is used.

To cope with this problem, a novel method for predicting the fault spectrum of a system was developed. This method is based on the conversion of the defect spectrum into the fault spectrum. The defect spectrum describes the number of defects per defect type. The defect types arise from a classification of the components and the production process into several classes. The defect spectrum can be predicted by knowing the composition of the system, and the defect rates of the components and of the production process. The conversion of the defect spectrum is based upon previous data. These conversion characteristics are independent of the technology.

5.8.1 Calculation of defect spectrum after manufacture

The defect spectrum of a board describes the average number of defects per board and per defect type. The defects are classified into types by the following aspects:

- component type
- manufacture step type
- repair type.

The classification into component and process types is because the different components and the different manufacturing processes have different defect rates. A further classification into repair types is only needed if for the same component type or the same process type different repair costs occur.

If the defect rates per component and the defect rates per manufacture step are known, the total number of defects per defect type, which forms the defect spectrum, is calculated as follows:

$$d_i = \frac{dpm_i \cdot n_i}{10^6}$$

where d_i stands for the number of defects for defect type i, dpm_i stands for the DPM rate (defects per million) for defect type i, and n_i stands for the number of elements of type i.

In ECOvbs we have defined the following defect types:

1. Component types: digital ICs, analog ICs, passive ICs, edge connectors, PLAs, RAMs, ROMs, microprocessors, ASICs, resistors, capacitors, bare board.
2. Manufacture types: pick and place, solder joint.

The defect types are defined in the cost models. Therefore it is very easy to add other defect types, or to remove any of the above.

5.8.2 Calculation of fault spectrum

The fault spectrum is derived from the defect spectrum by defining, for each defect type, how a related defect is distributed among the fault types. We call this definition the *conversion matrix*. The dimensions of this matrix are the number of faults (for each row) and the number of defects (for each column). The sum over each column of the matrix must be 1, as it represents the distribution of the population of defects. The fault spectrum can be derived from the defect spectrum by matrix multiplication. The defect spectrum and the fault spectrum are modelled as vectors **d** and **f**, and the conversion table is modelled as a matrix **C**. The fault spectrum **f** is derived from the defect spectrum **d** and the conversion matrix **C** as follows:

$$\mathbf{f} = \mathbf{C} \times \mathbf{d}$$

This method of deriving the fault spectrum, together with the calculation of the defect spectrum, allows a very accurate prediction of the fault spectrum before the system has been designed. This has been proven by making this prediction for computer systems at Siemens. The real numbers, which were derived when the system was in production, differed only by 2%.

Using the defect rates of the components and the manufacture steps for deriving the defect spectrum and the fault spectrum, has another major advantage besides the accuracy of the prediction: it allows the impact of different component or manufacture qualities on the manufacture yield to be calculated. This enables the mix of yield and fault coverage to be optimised for a target quality by optimising the total cost, and hence the optimum test strategy can be defined.

5.8.3 Calculation of defect spectrum for repair

In ECOvbs the repair costs are calculated not by using an average cost per repair but by determining the type of the defect to repair, and by using a repair cost per defect type. This method of repair cost prediction is more accurate than using an average value, because repair costs can be quite different depending on what needs to be repaired. For example, replacing a defective VLSI component is much more expensive than replacing a defective resistor. In addition, this method enables evaluation of the impact on the repair costs of improved manufacture quality, or the impact of alternative test strategies. There may be a difference in the repair costs for different test strategies if a defect - e.g. a defective component - is detected in a different stage of manufacture. For example, an income test of resistors may not only allow low quality charges to be returned to the supplier but may also significantly reduce the repair costs of the assembled board, if an exchange of components at board level becomes expensive.

In order to calculate the repair costs as described above, we must derive the defect spectrum from a certain fault spectrum. Based on the fault spectrum of the device under test before the test application, and the fault coverage of the test, we can derive the number of faults per fault type. But the repair costs depend on the defect type more than on the fault type. Therefore the number of defects per defect type needs to be derived from the faults per fault type. If the repair cost per defect type is given, the total repair cost can be calculated as follows:

$$tr = \sum_{j=1}^{n} d_j \cdot r_j$$

where tr stands for the total repair costs, d_j stands for the number of detected defects for defect type j, and r_j stands for the repair cost per defect for defect type j.

The defect spectrum of the defects detected, i.e. the number of defects per defect type, can be derived from the fault spectrum of the faults detected, the fault spectrum before the test and the defect-to-fault conversion matrix as follows:

$$d_{\det_j} = \sum_{i=1}^{n} f_{\det_i} \cdot \frac{f_{ij}}{f_i}$$

where d_{detj} stands for the detected defects of the defect j, f_{deti} stands for the detected faults of type i, and f_i stands for the number of faults of type i before the test. f_{ij} stands for the number of faults of type i before the test, which are related to the defect type j. This value can be derived by using the defect-to-fault matrix as follows:

$$f_{ij} = d_j \cdot c_{ij}$$

where d_j stands for the number of defects of type j before the test, and c_{ij} is the element in the jth column and the ith row in the defect-to-fault conversion matrix.

The values for f_{ij} are the same for all test methods, and therefore separate cost models, one per defect type, are provided in order to calculate these values.

5.9 THE TEST STRATEGY PLANNER

The test strategy planner creates a test strategy for the cost evaluator by creating the related cost models, creating a cost model connection file, which is itself a cost model, and verifying the applicability of the test stages and the consistency of the test strategy. The terms are defined below:

A *test stage* is a combination of a test method with appropriate test equipment, which is applied at a production stage defined by the user of ECOvbs. The production stage can be component, test cluster, board or system.

A *test strategy* is a combination of test stages which can be applied at several production stages. A test strategy requires the applicability of the test stages and the test strategy as a whole.

A test stage is applicable if all DFT and construction requirements of the test method and the test equipment are applicable. A test strategy is applicable if all test stages of the test strategy are applicable and if at least one DFT alternative per component type is available, which fulfils all DFT requirements of all test methods which are applied. A DFT alternative is an alternative implementation of a component that can affect testing. For example, a standard and a boundary scan version of the same chip would constitute different DFT alternatives. Different quality levels of components would also constitute different DFT alternatives. This would allow the user to examine the economics of using higher quality products in order to make savings in in-house testing costs.

A test strategy is set up by the user. The system lists which data may be entered, and checks the entered data for correctness. The data which need to be entered in order to define a test strategy are the following:

- match name and user friendly name of test strategy
- number of test stages

for each test stage:

- match name and user friendly name of the test stage
- production stage at which the test should be applied
- test method

- test equipment.

Once the test strategy has been entered, its applicability will be verified. If the test strategy is applicable, the user is prompted to decide whether the test strategy should be generated. The test strategy generation performs the following actions:

1. Selection of a DFT alternative per component and for the whole board. If more than one option is applicable, the user is prompted to select one.
2. Generation of the test strategy description file and the test strategy cost model.
3. Generation of the test strategy dependent cost models. This generation includes the calculation of several parameters:
   a. The calculation of the DFT alternative dependent parameters.
   b. The calculation of other summarising parameters such as the total number of gates, or the total number of digital pins.
   c. The calculation of the DFT alternative parameters per test stage. These values allow the cost per test stage to be calcuated. Further description will be given in the following paragraph.

In order to calculate the costs which are directly related to a test stage, the incremental costs of applying the test stage must be determined. These are the cost of test generation, test application, diagnosis and repair, and the incremental cost of design and production. The incremental costs of design and production occur if a test method requires a certain DFT method, which requires a different DFT alternative for some of the components or the whole board. These incremental costs are determined as follows:

- For each component and the whole board, the DFT requirements of the test method under consideration are removed.
- For each component the DFT alternative, which has the lowest component cost is selected.
- For the whole board the DFT alternative which has the lowest total component cost is selected. The cost per component is taken from the DFT alternative with the lowest cost.
- All cost model parameter values are derived.
- The DFT requirements of the test method under consideration are set and the user selected DFT alternatives are chosen.
- The incremental costs are derived from the difference between the cost model parameters before and after setting the DFT requirements.
- Each DFT dependent input parameter in the test strategy dependent cost models will be composed of the basic value - i.e. the value without any DFT requirements - plus the incremental value per test stage. The incremental value as the term of a sum is multiplied by a Boolean parameter, which allows the term to be switched on and off. By this method, all costs which are related to a certain test stage can be removed by setting the related Boolean parameter to zero.

The test strategy cost model includes for each test stage one parameter which is connected to all Boolean parameters of the other cost models, which are related to the same test stage. This allows the costs related to the test stage to be set or reset by setting this single parameter to one or to zero.

In addition to the test strategy dependent cost models, two files are generated. The first file contains a list of all cost models used by the test strategy. This list needs to be ordered so that a cost model which accesses a parameter from another cost model is listed after the related cost model. The other file contains all the set-up data of the test strategy which are entered by the user. This file can be read by ECOvbs to set up a test strategy. By modifying this file, the user can directly modify the related test strategy.

When a test strategy is set up and the cost models and other test strategy files are generated, the economics of the test strategy can be evaluated. Some sample results are described in the next section. Once more the reader should note that these are specific examples, and will not necessarily be true for different companies and products.

5.10 RESULTS

This section will provide the results of some sample runs of the ECOvbs system, and give the reader a feel for the way fault spectra and costs behave with varying test strategies. The results are specific to the input parameters and depend on the design, test equipment, company accounting procedures etc. They do, however, indicate the usefulness of such a tool, and show the relative merits of test methodologies.

The analysis was done on a large, double sided board containing surface mounted components, including a number of VLSI. The board is a complex computer board, typical of a mainframe system. The main board data are presented in table 5.2. The data are typical industrial values. Costs were calculated in ECU (European Currency Units), as this work was part of a European collaborative project. Most of the costing data, such as the component prices, are test strategy dependent, and therefore they are not listed here. An In-circuit test could not be applied to the board, because it is double sided, and the pin grid is too small. Therefore the alternative test strategies relate to functional test, boundary scan test, a prescreen test using a manufacturing defect analyser, and a system test.

Table 5.2. Main design data of computer board

Parameter	Value
Number of VLSIs	40
Number of RAMs	28
Number of resistors	800
Number of capacitors	100
Number of analog components	100
Production volume	5 000
Production yield	50%
Number of solder joints	15 000
Total design effort	353 weeks

Four strategies were evaluated:

- functional test followed by system test

- boundary scan, functional test, system test
- pre-screen (using a manufacture defects analyser), functional test, system test
- pre-screen, boundary scan, functional test, system test

The results are summarised in table 5.3 [Dislis93]. Figures 5.8 and 5.9 show the fault spectrum for the first two strategies.

Table 5.3 Results summary

Test strategy	Design cost (ECU)	Prod. cost (10^6ECU)	Test cost (10^6ECU)	Total cost (10^6 ECU)	Yield after test (%)
Functional test, system test	941 391	85.3	11.1	97.4	99.65
b_scan, functional test, system test	957 883	90.4	4.35	95.7	99.90
Pre-screen, functional test, system test	941 397	85.3	9.38	95.6	99.73
Pre-screen, b_scan, functional test, system test	957 883	90.4	4.93	96.3	99.90

The boundary scan test strategies have a higher final yield after test. This is achieved at a lower cost than functional test alone, despite the increase in the cost of boundary scan components. This is due to the ease of diagnosis that boundary scan offers over functional test. The boundary scan test eliminates a large number of faults which otherwise would have to be diagnosed by functional testing (an expensive process). The cost breakdown for the boundary scan/functional test/system test strategy is shown in figure 5.10. There is a small additional cost in the design and production phases due to the use of boundary scan. However, these costs are compensated by increased savings in the test phase, as can be seen in figure 5.11.

A pre-screen test is an alternative way of reducing the number of faults passed to functional test. It is effectively a cheap test stage to catch common faults. This is an effective strategy in itself, without the use of boundary scan, see figure 5.12. In fact, in this case there is not a cost benefit in adding boundary scan, but there *is* a quality benefit. Adding pre-screen to the boundary scan/functional test/system test strategy offers no additional benefits in terms of yield after test. Boundary scan is just as effective in finding the faults that pre-screen testing would identify, and the introduction of the extra stage simply results in a higher cost. The fault spectrum can be seen in figure 5.13.

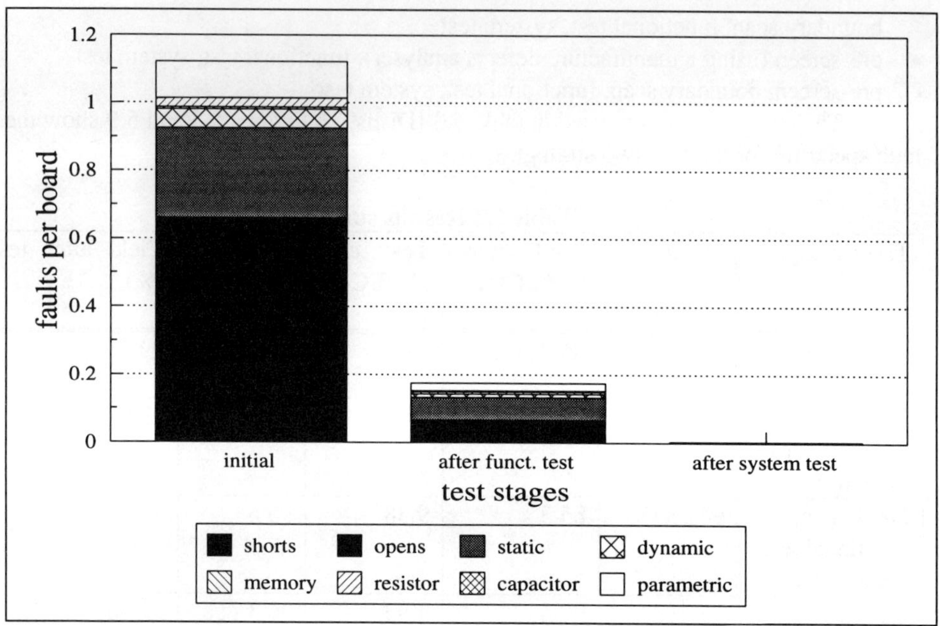

Figure 5.8 Fault spectrum for functional test/system test

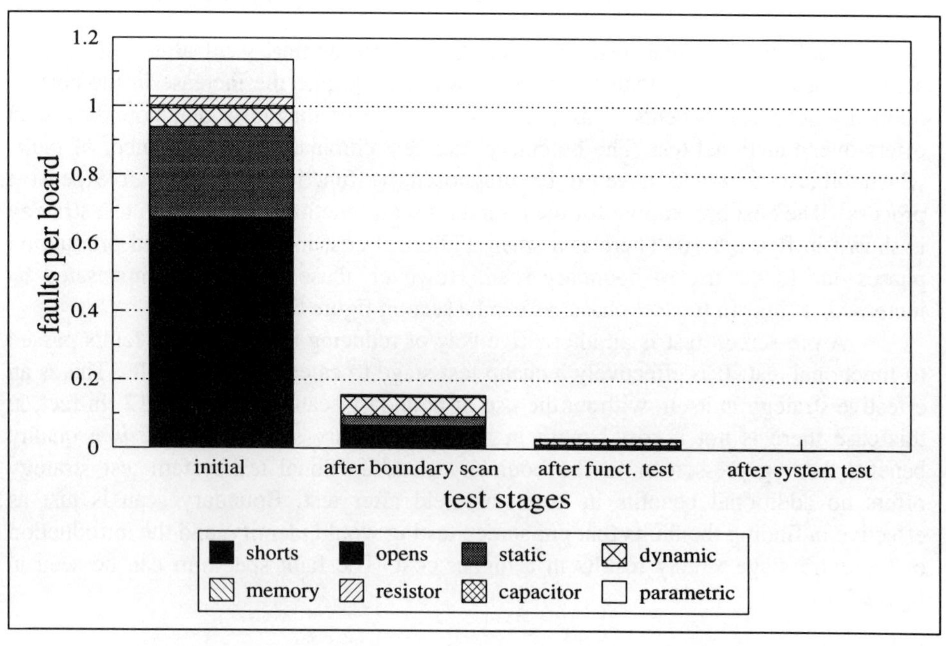

Figure 5.9 Fault spectrum for boundary scan/functional test/system test

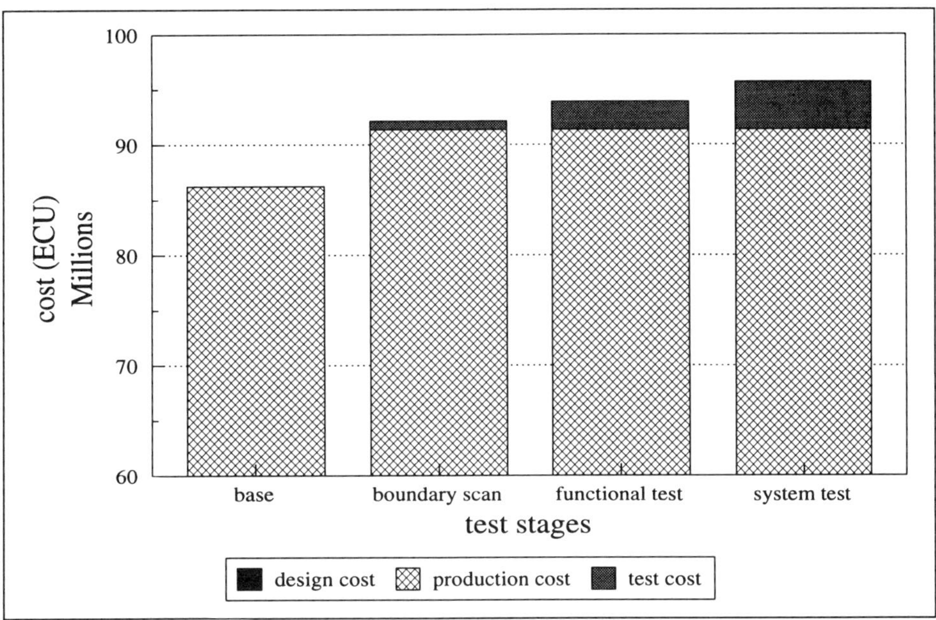

Figure 5.10 Cost breakdown for boundary scan/functional test/system test

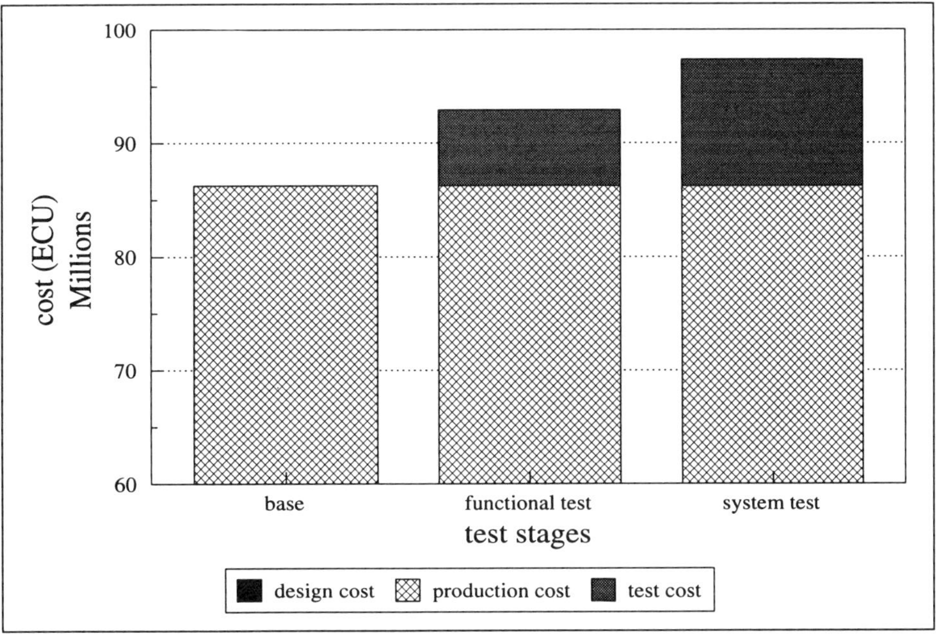

Figure 5.11 Cost breakdown for functional test/system test

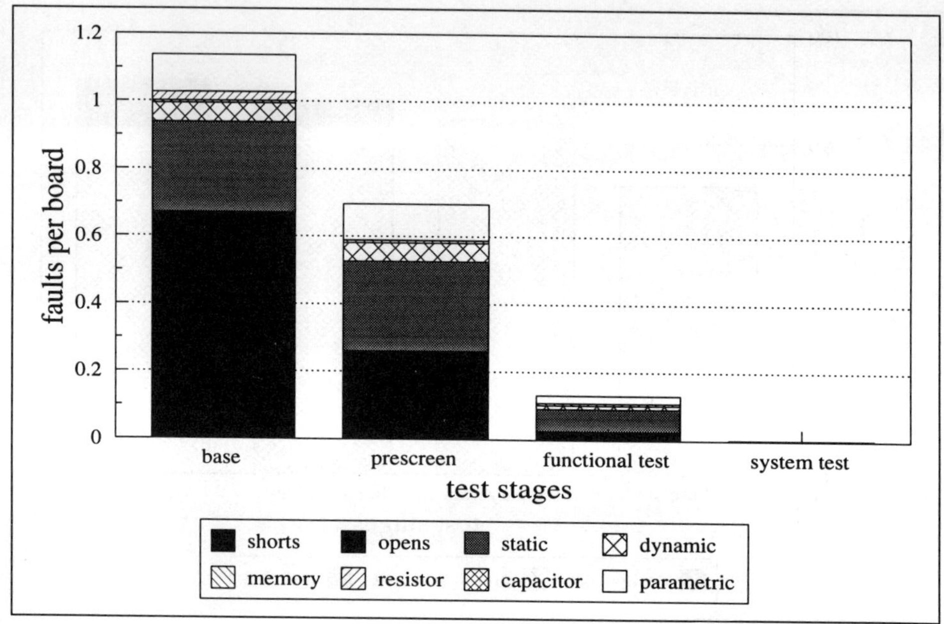

Figure 5.12 Fault spectrum for pre-screen/functional test/system test

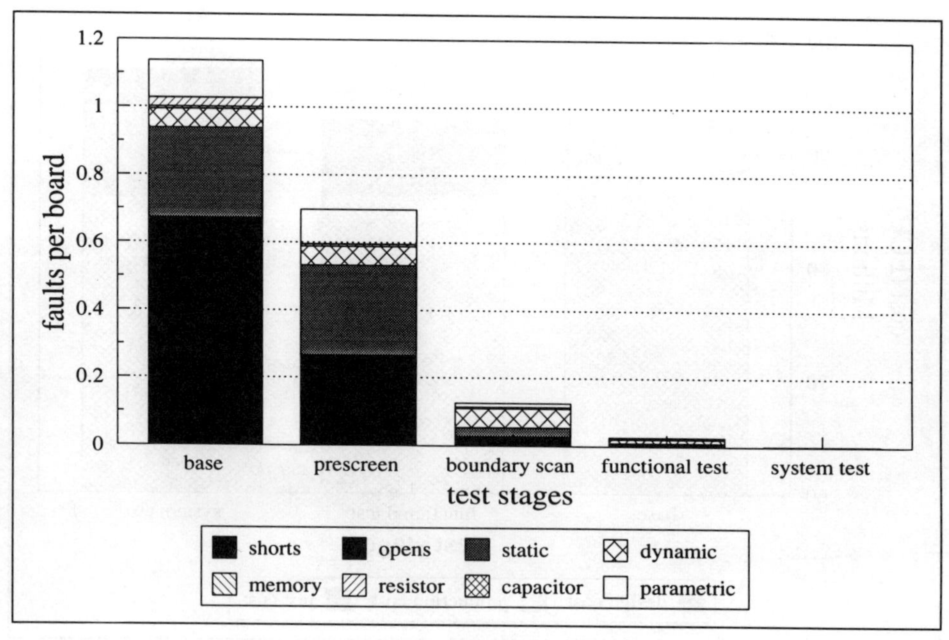

Figure 5.13 Fault spectrum for pre-screen/boundary scan/functional test/system test

The above discussion illustrates the point that simply adding extra test stages does not necessarily result in a higher quality product. This is a common misconception. If two test stages are discovering the same faults, then one is redundant. It is important to create a set of tests which will cover the fault spectrum adequately, and do so at a low cost. It was also shown that adding a test stage may increase product quality *and* reduce cost (by limiting the costs of subsequent test stages). Cost-quality trade-offs are not simple to evaluate, and in most cases it is up to the user to decide on an acceptable compromise. Remember also that some of the outcomes are not intuitive. The only way to make decisions on test strategy is by careful evaluation of all the parameters involved.

The above examples are only illustrations, based on typical values for one company. We would not expect them to hold in general. In fact, drawing up general guidelines on these matters is very tricky, if not impossible, as a change in only one of the parameters (fault cover of a particular ATE, for example) can alter the whole picture. One important aspect of using economics models and software tools based on these models is to understand their limitations.

6

Field Maintenance Economics

6.1 INTRODUCTION

To complete the life cycle cost analysis, this chapter will introduce the area of field test strategies and their economic impact. However, as seen in the earlier chapters, it is not possible to assume that they are mutually exclusive from chip, board and system test strategies. In developing a field test and diagnosis strategy it is not always possible to access or use existing DFT that has been inserted to aid in chip and board level testing. However, if consideration of field maintenance is given when selecting DFT methods many of these problems can be overcome, thus reducing the overall cost. Indeed, it is possible to add DFT in the design stages purely to improve the maintainability and serviceability of the system. The increased effort and cost at earlier stages can be paid back by the reduced cost in the field [Perone86, Fuller88].

The above analysis would obviously happen in an ideal world but all too often the pressures of the market place work against concurrent design procedures. Device designers often ignore a device's testability because the board designers are waiting for the device. A delay in delivery by the designer because the 'test department's problem' has not been fixed (e.g. the appropriate DFT added) is often not accepted. This happens at each subsequent stage of design and test. The field maintenance engineer is all to often left to solve the problems that could have been solved more cheaply and quickly earlier in the product's life cycle. If the problems are not diagnosable/repairable then *occasionally* this information is fed back to improve the quality of later versions of the product. The financial cost and impact on a company's reputation can be significantly damaged if a product recall is required. However, in general it is rare that the real cost of delaying the test decision is calculated and the information fed back to the right people to influence future products. Indeed, a company can be paying the cost for a product's poor quality and lack of design for maintenance for many years after the product has hit the market place.

Test economics at the field maintenance level is again company, market and product specific. As with the other test stages, economies of scale can heavily influence the testing strategy. For example, when there is a large product market, the manufacture of dedicated field test equipment and software can be an economical approach; for a small market large amounts of BIST or the use of a few highly skilled test engineers could be cost effective. The existing company infrastructure for field maintenance will also affect the choice. A large company that manufactures many different systems that have low down time costs, where there is a good level of technical support from a large pool of lowly skilled field engineers, with a mature and efficient logistic support environment, and well stocked repair depot sites, will have a very different viewpoint than company manufacturing a few high tech systems that have high down time costs (safety critical systems) and require highly trained field service engineers (or in some cases the original designers).

The situation is made more complex as the guarantee and extended warranty costs for a system are considered. A company manufacturing, testing and using its own equipment has a different perception of the cost and quality impact of field service than a company purchasing equipment and warranty from an original equipment manufacturer (OEM) or a contract service company. In general it is only the OEM that has the ability to consider DFT modification to the design to reduce field service costs. The contract service companies need to rely on more efficient logistic support, technical ability and test equipment to maximise profit by minimising costs. The profit to be made in the service industry is high and thus some areas are becoming highly competitive.

This chapter is aimed at improving the cost awareness of field maintenance procedures. It is important that the information presented in this chapter is not applied to specific examples within a particular environment without careful consideration.

6.2 THE COST OF FIELD SERVICE

In 1990 the estimated US service market stood at $46 billion. It was further estimated that 26% of the revenue and 50% of profits are generated by service centres of equipment vendors. Obviously, if the service warranty is a fixed price contact then the more reliable or cheaper to diagnose and repair a system the greater the profits for the service centres. If not, then it would seem to indicate that the more unreliable the product the greater the service centres' profit. However, there is strong evidence that the reliability of the product can influence its repeat sales and indeed the sales of all the company products. It can take many years to recover from a reputation for poor product quality. Many maintenance contracts in the industrial sector can carry very high down time costs, and it is highly unlikely that the down time penalty associated with a field failure is a fixed price for the period of the warranty.

Another factor to consider is the perception the user has of the quality of service obtained from a service centre and not the quality of the product. A user might well prefer a less reliable product (cheaper) if strong guarantees can be obtained for the quality and speed of field service. A consumer is likely to return to a company that has a policy of replacing an item if it fails within a guaranteed time. The consumer is now less sensitive to the reliability of the individual product. A variation on this idea is to ensure that consumers have a replacement until their own product is repaired. Both the above methods have been shown to be cost effective in reducing life cycle costs for a product and reducing time to market while still maintaining customer satisfaction. Such approaches need very careful consideration before being accepted as the risks can be high.

The above permutations can be considered part of a global field service test strategy. Indeed this corporate decision should be a prime consideration when developing the design and test specifications for an individual product. Obviously, to generate an economic model that will enable such variation to be analysed for the many different companies, contracts and products is difficult, if realistically possible at all. However, detailed analysis can successfully be performed inside a specific and mature environment for a particular product. This has proven to be very valuable in assessing the *real* cost of field service and the potential for profit. Indeed it could act as a tool to guide a company in making the correct corporate decisions for future products.

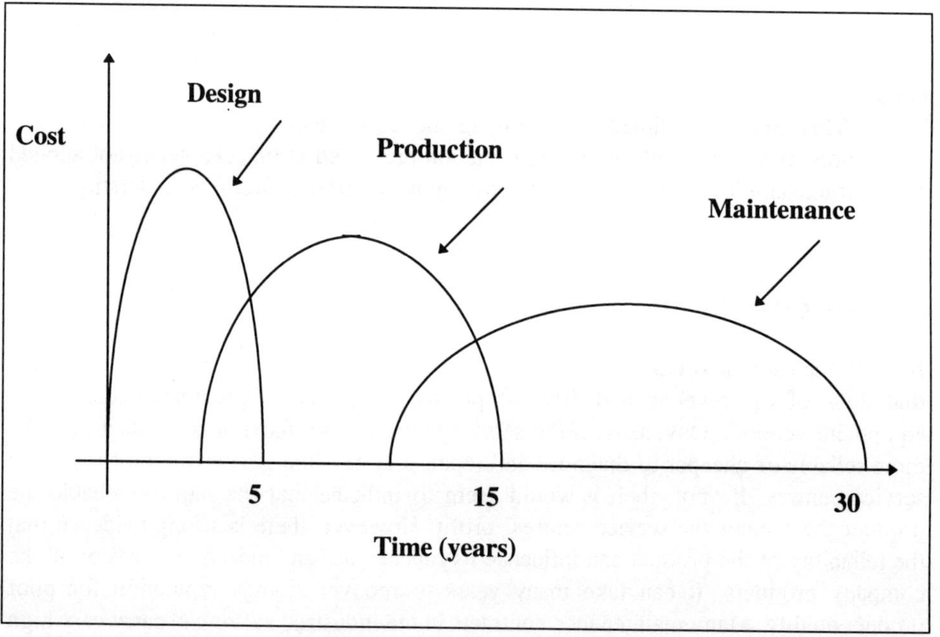

Figure 6.1 Life cycle costs

Published studies of life cycle cost are few and far between but some studies from a military environment have indicated that the field service cost can approach 60% of a product's total life cycle cost, while production costs approach 30% and design costs only 10 % [Deway86]. This was for a product with a 30 year life in a military environment. Figure 6.1 shows the life cycle cost for this particular case. Commercial products with an expected working life of two years might be very different. However, figure 6.1 does indicate the scale of the potential savings that can be achieved by reducing maintenance cost. Indeed, in some cases the cost of repairing old equipment can be higher than replacing it with new equipment. A US army study has shown that costs could be reduced by 25% if 70-80% of the items it currently repairs were discarded [Murray88].

Analyses of other industries have also indicated the high cost of service contracts. For example, it has been quoted that 35% of a user's total computing costs derive from computer service costs [Fink84], and 15-30% of a computer manufacturer's profits come from field service contracts. For example, the cost of extended warranties on personal computers needs to be carefully considered. Indeed it is often not cost effective for large educational organisations to have extended warranties for all their personal computers. If one fails and can not be repaired cost effectively in-house then a new machine is purchased. This can be partly due to technical support and possible low down time costs but also the short working life of the PC before it is considered obsolete. The domestic market can be very different. For a domestic video recorder a four-year warranty in the UK from a large retail chain can cost around £100. This is for a machine with a purchase price of between £250 and £400. Extended warranties for future years become significantly more expensive. The average repair cost for such an item obtained from an independent source is around £40 labour and £49 parts, and on average 70% of the repairs require new parts. Most of the machines repaired were over three years old and required new motors, recording heads or loading carriages. Perhaps the consumer is purchasing peace of mind by taking out a warranty rather than taking the probabilistic view of the likelihood of a failure occurring within the first four years and the repair cost being greater than £100.

6.3 TYPICAL SERVICE PROCEDURES

In fact it is almost impossible to have a single typical service procedure. However, in this chapter a simplistic view will be taken to enable the various cost areas of the service industry to be discussed. Figure 6.2 shows a simple field service procedure. It is a two level strategy (i.e. there are two possible locations where failures can be corrected). Different industries and companies will adopt different levels of field service. Indeed, the US army during the Vietnam war adopted a six level strategy for its equipment [Murray88, Hughes89].

- Level 1: Soldier
- Level 2: Organisational
- Level 3: Direct support
- Level 4: General support
- Level 5: Depot
- Level 6: Manufacturer (OEM).

However, the current trend is towards reducing the number of levels of repair. As the complexity of systems increases the cost of repair centres increases and the diagnosability and repair of faults becomes an expert task. A currently adopted strategy is a three level strategy: unit (or field), intermediate, and depot (or OEM). The US Air Force adopts a two level strategy containing a field repair and depot (OEM) repair. This is due to the location of its systems, complexity of its systems and the need for high levels of quality assurance. In the digital-avionics industry the removal of the intermediate level can produce savings estimated at 45 % [Bryant89].

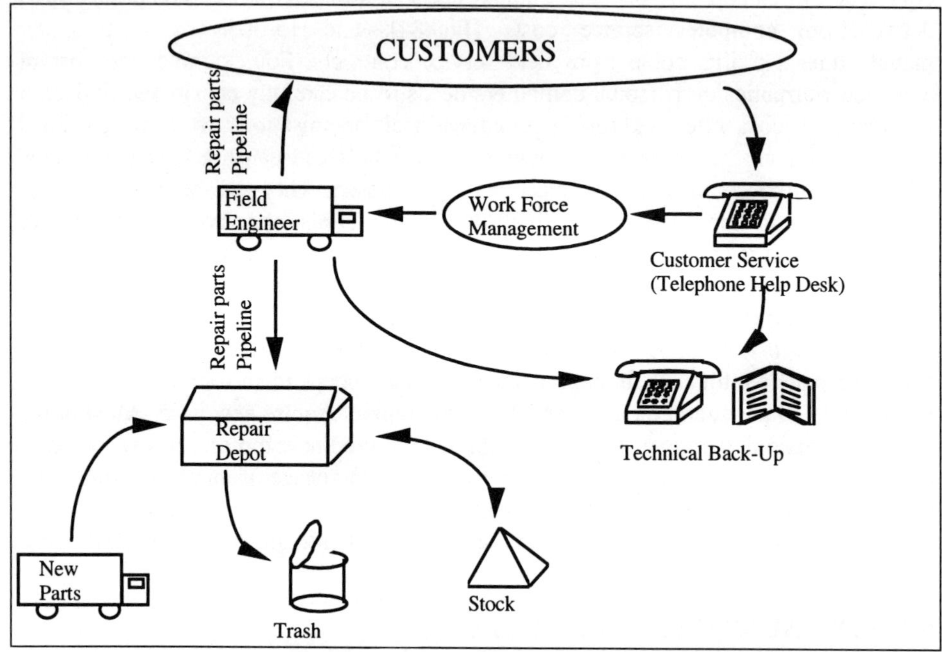

Figure 6.2 Typical field service procedure

A simplified scenario for a large company is as follows. The first stage of a service procedure is initiated when the customer service department receives a call from a user who reports a fault. At this stage an attempt is made to diagnose the type of fault. In some cases a reported fault could be a user error. The accurate recording of the failure symptoms can significantly improve the quality of service and cause a reduction in the overall cost. The efficiency of this process can be governed by day to day variations in human interaction. The cost effectiveness of such arrangements is hard to model. In addition, there is often technical back-up available to assist the large number of low skill customer service employees. In smaller companies where fewer faults are reported the skill level of the customer service department can often be increased as in some cases customers will contact the technical back-up directly. The skill level of the customer service department can affect the quality of the initial diagnosis procedure.

The initial diagnosis is used by the work force management to prioritise the service calls and detail the work schedules for the individual repair persons, thus minimising the response time and ensuring that the repair person with the correct skill level, documentation and spare parts is dispatched.

The first level of repair is carried out by the field engineer, whose test equipment and repair tools are often simple and inexpensive. However, this will depend on market conditions. Often a cook book approach to the maintenance and repair procedures is followed. Should the engineer have difficulties she/he can often contact the technical support for help. If the diagnosability of the fault and its repair are beyond the scope of the field engineer dispatched then a repeat visit is required by a more qualified and better equipped engineer. When a field engineer is not responding to a reported fault call she/he will be performing regular maintenance checks and repairs.

The increased level of integration and hence complexity has led to repair personnel carrying out a diagnosis and replacement strategy, with the faulty field-replaceable unit being returned to the depot (or OEM) to be repaired or discarded. In some cases where there is a need for very high repair quality, any board removed from a system has to be returned to the depot before it is returned to the field. With such a procedure it is essential for the field repair person to diagnose the faulty board accurately. It is often the requirement at the design stage that field failures can be diagnosed down to a single field-replaceable unit. However, this is rarely achieved for 100% of all failures unless significant levels of BIST or complex ATE are used. It should be remembered that not all field failures are a function of hardware faults. Indeed, in several situations that have been examined hardware failures can contribute to less than half of the service calls. Furthermore, it has highlighted that poor customer training can result in 4-10% of calls. This is a contributing factor to the growing number of calls where no fault can be found.

All boards returned to the depot are then checked and repaired if it is deemed cost effective. Again, if high levels of reliability are required it will be necessary to assign a unique identification number and to maintain a repair record for each board. To ensure reliability there is often a maximum number of repairs allowed per board. Any board found faulty by the repair person but deemed fault free by the depot often requires significant levels of test before it is returned to the spare parts store. Any questionable part is discarded and a new one ordered to maintain stock levels. Our analysis of a non-automated field service industry has shown that about one third of all returns to the depot are in fact good units.

6.4 MODELLING THE CAUSES OF HIGH FIELD SERVICE COSTS

When developing a model of the required complexity it is important to understand the areas of major cost concern in the industry. The need for detailed models will thus be a function of the sensitivity of the area and the room for possible improvement. In some cases a lump sum cost could be all that is required. However, the difficulty or ease of obtaining data should not be used as a reason for producing a more complex model. Indeed the process of data collection might highlight the need for a company to obtain certain data to improve future corporate decisions. The level of human interaction and its variety in the field service industry will make the data collection for detailed modelling difficult and time consuming. In this section we will discuss some of the areas that need to be considered for modelling and data collection in the field service industry.

6.4.1 The depot

The economics modelling of the repair depot can be achieved by making use of the basic models that have been used to model the cost of a board level production test stage. The depot will often have the same test facilities as the OEM. As mentioned above, the board is the most common field-replaceable unit for complex electronic systems. Thus the ECOvbs system can be used to consider the depot cost of performing a repair. Indeed, it is possible when using the ECOvbs approach to analyse the impact that the manufacturing out-going quality levels have on the total repair cost at the depot. Thus the direct result of the different board test strategy options can be followed through to observe the impact on the repair costs at the depot. Two other main depot cost areas that should not be ignored are maintaining the stockpile of spare parts and the cost of discarding and replacing parts. Indeed, the amount of company assets that are tied up in the stock of spare parts should not be ignored. It is often cheaper to replace faulty obsolete equipment with new models if parts are no longer in manufacture than to stockpile a large number of parts.

The following is a brief summary of parameters or models that are required to calculate the depot costs, quality/reliability of each repair and general quality of service:

- labour costs/number of staff
- staff training costs
- average number of replaceable units handled
- required utilisation factor
- cost of maintaining service documentation
- ATE cost
- required throughput rate per replaceable unit
- required stock level per replaceable unit
- stock level in pipeline at depot (replaceable units waiting or being tested and repaired)
- purchase cost of the replaceable units

For each replaceable unit

- its reliability
- likelihood that it unit has been repaired more than a pre-defined number of times.
- effect on its reliability due to repair
- spares cost for each part of the repairable unit
- required stock level per part
- percentage of replaceable units repaired
- percentage of replaceable units that are non-repairable
- percentage of replaceable units tested and no fault found

- diagnose time for each fault classification
- repair time for each fault classification.

6.4.2 Logistic support

The harder areas to model are those that involve greater human involvement. Indeed the level of human interaction in the field service industry, along with the delocalisation of the field testing activity, are the main reasons for high field service costs. It has been reported that the cost of the service organisation can be in the region of 65-75% of the service industry's revenue [Blumberg90]. It is possible to generate complex logistic support models and compare the efficiency of different logistic structures and procedures. However, at the other extreme, in most cases it is possible to consider logistic support as a lump sum cost. In many cases the lump sum cost can be effective due to the large percentage of costs related to staff employment. If a lump sum is used then the scope for improving the quality of the service should not be ignored. However, it is difficult to generate accurate predictive measures unless a lot of time and effort is spent on the process. Our initial studies have shown that simple procedures can generate a significant improvement in the quality of logistic support with little or no cost impact. A well managed and motivated work force is essential, and so is encouraging staff to be responsible for their decisions. In addition, forcing staff to make a decision on the diagnosis of the fault on the initial contact with the customer can reduce the number of expensive repeat visits. In busy periods 60% of reported faults were logged incorrectly or left undiagnosed by the initial contact, this was a significant improvement on the 40% average for other periods. One such policy improvement can be to allow only the technical back-up personnel to log a fault as unknown. The improvement in training and the generation of better fault logging questionnaires and procedures can help.

With such a large dependence on human interaction, staff prediction models are required to calculate the speed of service or service quality. These models can be very similar for all the labour forces (field engineers, depot staff and logistic support staff). The models can be very simple; at the limit they will be a factor of the total number of staff, total number of activities, and the required wastage (utilisation factor). The ultimate complexity of the models will be a function of the size and sensitivity of the service industry. In the case of the depot, at the extreme, a poor throughput at the depot, due to poor staffing, can increase the need for a higher level of spares. Alternatively, it can produce long delay times while service staff wait for a board to be returned or for spare boards to become available.

For management to predict the optimum level of staff it is essential that there is an accurate prediction of the number of field failures or service calls made. In addition, reliability figures need to be categorised into the different types of systems and field failures. To obtain a greater accuracy the prediction can not only be based on mean time to failure information, but seasonal and weekly fluctuations need to be monitored.

If we consider again the domestic TV or video recorder there is often an increase in service calls during public holiday periods (Christmas, Easter, etc.) and during or immediately after weekends. Also, climatic changes can result in fluctuations in the number of service calls. In addition, some companies might experience fluctuations during the day, for example office staff might report more faults with office equipment when they are switched on in the morning or after lunch. Figure 6.3 indicates how the number of calls might change with time.

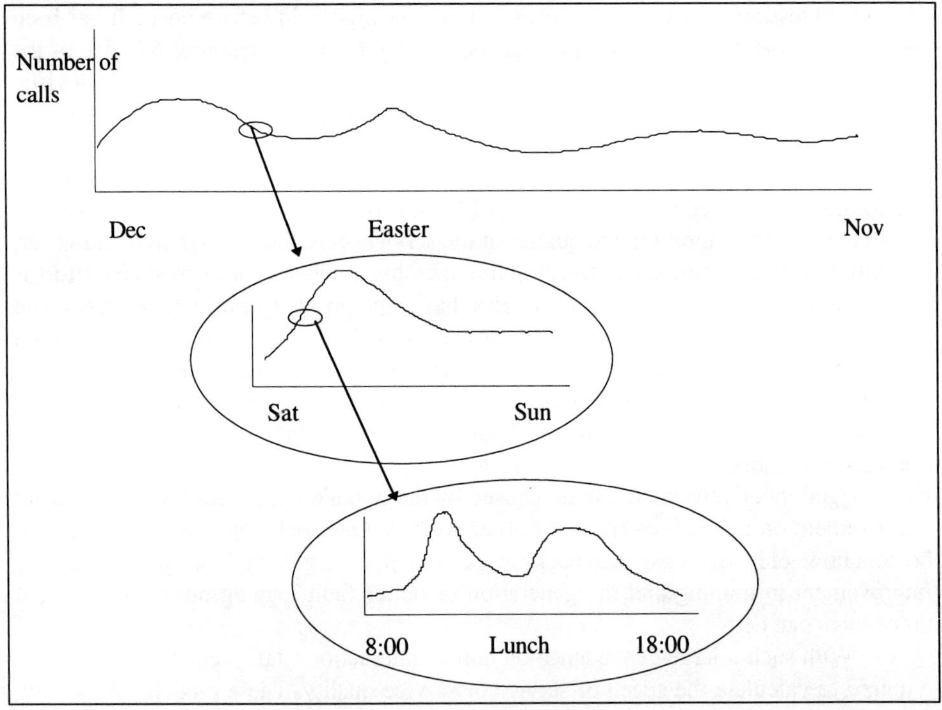

Figure 6.3 Number of service calls

6.4.3 Field repair cost

The cost per field repair can be calculated from knowledge of the field diagnosis and repair time for different failures and the cost of replacement parts. Each field engineer will have a slightly different profile, but due to the sensitivity of the modelling and data collection procedure it is often possible to divide engineers into groups of different skill levels. Thus the diagnosis time and repair time can be associated with a particular engineer's skill or the field test equipment carried. The identification of the number of field engineers required in each skill group and type of field test equipment carried could be considered part of the global field service strategy plan. There will be a set of optimal solutions to this problem and this mix of solutions will change as the product and market conditions change.

As mentioned in the previous section, a valuable aid in predicting the human resources necessary for field repair is the knowledge of the likely number and distribution of field failures. This will affect the number of field engineers required. However, the number of field visits is not equal to the number of field failures. Our analysis has shown that in the order of 20% of field visits are repeat visits to service the same fault. There is also a percentage of faults that are never diagnosed. These could be due to a temporary failures that cannot be repeated or a user reporting a fault in error. In addition, field engineers are often required to perform regular maintenance of systems and this needs to be costed.

For each field engineer the company needs to make a significant investment in spare parts, test equipment, transportation, training costs, etc. The pipeline cost (i.e. the cost of spares and faulty units held outside the depot) is not insignificant. The spares pipeline, in some cases, can tie up 5% of a company's assets [Comerford82]. Approximately 80% of the value of the service pipeline is returned through the depot repair operation annually [Blumberg90]. As the complexity of replaceable units increases, the cost of the pipeline will rise unless new practices for repair are considered. The more in-field repairs that can be undertaken the more the size of the pipeline will be reduced. However, this has an impact on the skill and training of the repair personnel, the number of staff required and the investment in ATE for the field engineer. In addition, such a strategy might have an impact on the quality of the service. In a well tooled repair depot it is possible to obtain consistency in repair quality which is monitored easily and corrected quickly.

A summary of the parameters that are required or models needed is given below:

- number of field engineers
- average number of calls handled
- average response time to call
- average diagnose and repair time
- average time wasted per visit
- utilisation level of staff
- number of multiple visits required per fault reported

For each skill level

- cost of replaceable units carried
- average number of replaceable units waiting to be returned to depot for testing (pipeline size)
- ATE costs
- cost of documentation carried
- staff training costs
- transportation costs
- percentage of faults not found

- diagnose time
- percentage of hardware faults
  - diagnose time for each replaceable unit
  - repair time for each replaceable unit
  - cost of repair for each replaceable unit
- percentage of software faults
  - diagnose time
  - repair time
  - cost of repair
- percentage of other faults
  - diagnose time
  - repair time
- cost of repair.

6.4.4 Down time costs

The down time cost is another factor that can drastically affect the cost of field service, particularly when safety critical systems are considered. Recent disruptions in finance companies due to terrorist action and system failures have resulted in significant losses (billions of dollars). Ten years ago the down time cost of an airline reservation system was around $24 000 a minute [Warwick84]. Due to the increased amount of air traffic and inflation, costs would be higher today. Even if the down time cost is small, as is the case for most consumer products, long down times do indicate a poor service quality, thus affecting system resales and future maintenance contracts.

It is hard to generate a general model for the predicted down time cost and its effect on the future marketability of the product. However, when clear penalty clauses are incorporated in the service warranty then the real cost risk can be assessed. Accurate methods to predict product reliability and repair times are required to ensure that the service company does not lose money when setting the terms of such contractual agreements.

In some cases the response time of the field engineer to the field failure needs careful consideration. It can take the engineer longer to respond to a call than it does to fix the faulty equipment. This is partly due to the efficiencies of the logistic support and the human resource levels for field engineers. Also, it should be remembered that the distribution of the faults that are reported is never constant, but varies with the time of day, seasonal changes and unforeseen occurrences.

6.5 REDUCING THE COST OF FIELD SERVICE

In this section we will briefly discuss how field service costs can be reduced and thus identify areas where detailed models could be developed to aid in the analysis of their cost effectiveness.

6.5.1 Repeat visits

As mentioned earlier, the number of repeat visits can be significant. In many cases a repeat visit incurs a cost that could be avoided. Repeat visits can result in increased response time to other failures, thus increasing down time cost and affecting customer perception of the quality of service. They can lead to an increased number of good boards being returned to the depot, thus impacting depot testing costs and also the size of the pipeline. The causes of a repeat visit can be attributed to four main factors.

1. Poor diagnosis by the initial fault reporting mechanism. This can be considered due to badly equipped or trained staff and may result in an engineer with the wrong experience being sent or the wrong spares taken and thus the need for another visit.
2. Poor repair of the faulty equipment. This is often a result of the lack of automation in the field service industry. The high reliance on humans results in the process being prone to error and inconsistencies, especially if poorly skilled engineers attend repairs that are beyond their training.
3. Inadequate/out of date field service documentation. It is not unknown for a service engineer to arrive at a site expecting to see one piece of equipment and to be faced with a completely different one or at least a different version.
4. Poor initial design, making fault location difficult and leading to incorrect diagnosis and repair.

6.5.2 Design for maintainability

Design for maintainability can cover a host of different subjects. It does not always have to mean an increased level of DFT, BIST or redundancy in functional design. Very minor improvements to mountings or the position of unreliable equipment can reduce repair time. Increased protection or better test specifications for unreliable equipment are also beneficial. However, any design concept that can remove the skill level or sensitivity to human interaction in the field service industry will have its benefits. The cost implications will need to be considered to ensure a significant reduction in life cycle cost.

6.5.3 Improved maintenance procedures

An obvious way to reduce the cost of field failures is to reduce the number of field failures. This can be achieved by increasing product quality and improving maintenance procedures. The Chrysler Auto Corporation, in order to increase the production quality and remove the costly downtime due to unexpected failures in its production line, instituted a practice of shutting down all equipment every 24 hours to perform complete preventive maintenance procedures. This simple approach proved to be highly beneficial. Indeed it is common for items such as photocopiers to inform the user when it is time for a maintenance check. These tend to be simple procedures based on counting the number of copies, but the idea holds for a complex system using BIST structures to monitor performance or the length of the system's operational time.

6.5.4 Remote monitoring

Poor logging or poor diagnosis of the fault by the initial fault logging procedure and poor logistic support can result in the wrongly skilled engineer being sent to repair a fault with the incorrect documentation, test equipment, and replacement parts. The use of remote diagnosis can have a significant effect on the number of repeat visits along with a reduction in the number of initial visits required. Remote monitoring is common practice for several large mainframe companies. The on-line monitoring of failures aids them in predicting maintenance schedules and improving the systems test procedures. AT & T have developed a system to perform remote diagnosis and maintenance monitoring that is estimated to produce a reduction in service calls of 50%. An estimated reduction of 10-40% in service calls for hardware faults could be achieved by remote diagnosis. Indeed 90% of software faults could be corrected remotely [Blumberg84].

It is often the case that large amounts of failure analysis data are collected via on-line monitoring but very little information is fed back from the field maintenance department to the original design and test departments to improve design for maintenance and system quality.

6.5.5 On-line documentation

On-line documentation is a cost effective way to overcome the problem of a service engineer not having the correct or most up to date documentation. This could reduce the number of repeat visits, training time, the need for on-site specialists, the number of good boards swapped, the pipeline inventory, and repair time.

The lack of automation and thus high dependency on human interactions causes many inconsistencies in service. Indeed services are prone to error. Any incorrect diagnosis not only results in repeat visits and down time but also an increase in the number of parts being sent back to the depot for repair. The cost of ensuring that a board reported bad by a field engineer is in fact good is not insignificant. There is also the consideration of the degradation in reliability of the board to be considered.

6.6 FUTURE TRENDS IN FIELD SERVICE COSTS

The continued rapid explosion of field service costs should be expected for several reasons:

1. The complexity of systems and boards is increasing and will increase the cost of the replacement components. This will have an impact on the spares and pipeline costs.

In addition the effort required to make the systems field-testable and diagnosable will need to be increased.

2. The process of repairing boards is becoming more complex. The use of multi-chip modules and other such technology improvements increases the cost of a repair. There is a greater need to ensure quality of repair for such systems otherwise their mean time to failure becomes intolerable.
3. There is a rising level of system customisation. Many equipment installations are tuned to meet specific customer requirements. This could be from both software and hardware modifications. The variety of installation can cause problems in maintaining documentation up to date. This could result in a need for an increased level of training.
4. The working life of systems is decreasing, increasing the amount of training required but also reducing the amount of on the job experience gained. The field engineer's learning curve can begin to play a significant role in the repair time.

It will thus become more important to consider design for maintainability and automated field test equipment to over come these problems.

6.7 DATA VALIDITY

The high reliance on human interaction in the field service industry makes the data collection and validation procedure particularly difficult. In many cases large variations in the data collected become apparent. When there are considerable and unexpected fluctuations in data it is important that the source is carefully examined. Within a particular study that we performed, anomalies were most noticeable from data received from field engineers. For example, sometimes a large number of boards would be replaced in systems, and no fault would be recorded in any or most of the boards when returned to the depot. Such anomalies, in this particular study, had a significant effect on the efficiency of the field repair process and a large impact on costs. A major requirement when collecting data is thus to never use data unless their true origins can be verified.

6.8 SUMMARY

This chapter has highlighted some of the major areas of cost in the field service industry. The amount of improvement and saving in each area will of course differ among companies and industries. However, it is an area that needs consideration and is becoming more important to allow designers time to address some of the problems that result in high field service costs. The problem will not go away and cannot be passed from a field service department to anyone else. More importantly, economic considerations are vital, as the field service industry has a large money making potential.

7

Conclusions

7.1 THE NEED FOR ECONOMIC ANALYSIS

This book has dealt with some of the issues related to the design and testing of complex circuits, and specifically with the economic effects of test decisions. The use of economics models was demonstrated, and the authors attempted to provide the reader with an understanding of the construction of such models. This understanding is essential, so that models can be tailored to reflect the real operation and play a part in evaluating decisions. The need for such an economic analysis of test issues can be supported from several viewpoints:

- the designer
- the test engineer
- management.

Designers traditionally have remained unconvinced about the merits of design for testability. Although this is beginning to change, it is a fact that DFT is not taken up to the extent that one would expect given the arguments of increased complexity and increased quality requirements. Some designers maintain that their priority is achieving performance targets and pushing the greatest possible functionality into the tightest possible space. There is also a fear that rigorous testability guidelines limit creativity, but the main point is that testability is not seen as the designer's responsibility. This is an interesting situation. Designers will take into account manufacturing tolerances that will affect functionality and performance, e.g. fan-out/fan-in, gate delay distribution and the like, but not manufacturing defects and the required detection methods.

However, from the test engineer's point of view, there is a problem to be addressed, caused by, on the one hand, design errors, and on the other by manufacturing defects. Testing to a high fault cover can be very hard with current circuit complexities and pin-to-gate ratios. It seems that in this scenario, the design department is creating problems for the test department, which in turn would require a redesign, thus expending time and money. The test department is not usually involved in the specification of the product.

Conversely, one more interesting variation is where a test department, in an attempt to increase its standing within the company (empire building?), may suggest, with appropriate justifications, that investment in the next generation of mega-tester, costing, of course, several millions of dollars, is the only way to proceed. This may temporarily ease the testing problem, while increasing investment in the test area. Management may now, for the first time, become aware of test cost issues and begin to consider the use of DFT. A thorough economic analysis might prove that investment to be unnecessarily expensive.

Either way, *everyone* associated with the product should be aiming to provide a manufacturable product that a customer will want. The problem is often one of communication (lack of), and is now seen as one of the main factors of slow development cycles. Concurrent engineering methodologies attempt to resolve such problems by breaking down departmental boundaries and involving everybody in all stages of development. In a concurrent engineering environment, the test engineer would make testability part of the initial specification of the product, rather than have to treat it as an after-thought.

The type of economic analysis discussed in this book fits into the concurrent engineering philosophy. Economics analysis and economics models are tools which can be used across departments, and provide a financial basis for decisions which otherwise might be taken subjectively - an economics model, as suggested, can be used to compare objectively the 'value' of a very wide range of disparate quantities. From the point of view of management, they can be used for long term planning as well as for early test planning for a particular product or product line. Again, arguments are backed up by hard financial facts, and therefore design for testability planning can be tackled in the same way as other decisions - on the basis of economics. The important factor here is that all relevant information should be taken into account when building the economics model - an evaluation of DFT in terms of silicon area alone would almost always be unfavourable. A realistic evaluation would include a financial analysis of the effect of simpler testing. Such numbers can be taken in several ways, either to show that use of DFT is not going to cost as much as was thought, or to show that not implementing DFT will result in enormous extra costs to the company through investment in test generation, ATE and field test requirements. Alternatively, the cost analysis may show that using DFT in certain designs with given production runs is not appropriate or necessary. An extra benefit of the cost analysis exercise is an educational one. By analysing the costs involved, the company may realise the need to educate its designers in the use of DFT.

This book has been based upon the experiences of the authors who have been looking into test economics in association with a number of major electronics companies for over ten years. The benefits to these companies have been obvious and significant, causing them to make changes in basic design strategy and the implementation of test environments.

The results that have been referred to should be of direct use and interest to designers, test engineers and management. But it must be remembered that the results, while pertaining to *real* situations in *real* companies, can only be applied to the specific situations analysed. However, the data does show interesting trends and where the benefits of such analyses have been found to be. One aspect that resurfaces is not only the usefulness, but also the usability of the models. The number of parameters involved can be daunting, and creating a good model from scratch can be time consuming. However, once the model is created, good organisation of the parameters should ensure its usability. For example, parameters which relate to the company and are design independent do not need to be updated by the user. Clear presentation of results is also important from the user's perspective. It is essential to think carefully about the user interface of models such as these, which will potentially be utilised by several departments. Ease of use is very important, as users will not accept an unwieldy piece of software. A well structured model on the other hand, where different strategies can be evaluated with ease, will be accepted with the minimum of trouble.

7.2 PROBLEMS WITH ECONOMICS MODELLING

The issue of using cost as an arbiter in decision making during design should not be an issue. Despite the contents of previous chapters, it must be reiterated that when it is considered that approximately 70% of life cycle costs can be determined at the design stage, costs should be very much to the fore in all designers' minds at all times.

It can be put in many different ways - in any product where costs are an issue, a designer *must* have an understanding of the costs in order to be able to make intelligent decisions about the inclusion of design features. If the appreciation of costs is not apparent, then potentially irrational decisions will be made.

But in essence, to say that cost is a topic that should be addressed is not the only aim of this book - it is to also suggest that cost estimation as a tool is a worthwhile aim. Indeed, it might also be suggested that cost estimating should be a fundamental engineering management tool in the design environment, and crucial to good project management [Kitchenham90].

Having created or agreed the need for cost estimation within your own organisation, it is important to be aware of the pitfalls associated with its usage.

7.2.1 Creating economic models

It will be very easy to create an economics model that is used in anger on real projects, and yet rely too much on the results of the model. To fail to appreciate the limitations, potential or otherwise, on the methodology will certainly cause the project to suffer. Any simulation or representation of reality has its drawbacks which the serious practitioner should be aware of.

One factor to appreciate is that costs will typically be an emotive topic for everyone to want to make a statement about, and to have an opinion on. This can be dangerous, but the problems can be mitigated by ensuring that the economics model presented to management for approval includes absolutely *everything*. The benefit of doing this ensures that the model creator will leave no stone unturned, which in turn will reduce the opportunity for potential detractors to find problems. Detractors will otherwise be found not only among those against cost models in principle, but also among those with vested interests (personal empires?), e.g. the battle between the design team and those involved in production, test, field test and marketing. A simple model, while having obvious benefits from the point of view of initial creation and maintenance of the data, can be seen to be potentially 'dangerously ignorant' [Kitchenham90] - simplicity should only be used where simplification has been shown to be possible, i.e. after the initial model creation.

Is there a short cut to achieving a cost model? It is possible/likely that a commercial model might exist, but the likelihood of its being directly applicable to any given situation is limited. Even a model created for one situation and applied to another that is *perceived* to be similar will be likely to have important parameters derived using different assumptions. Basic data will obviously be different and new data sets collected, but even accounting procedures can be the cause of differing output figures from otherwise identical situations. Cost estimating does need to be based upon individual circumstances.

Having created a model, a serious decision must be made about who should use it! For example, should it be the project manager, or an independent team? Perhaps the project manager would be the correct person to use the model as he/she will have most to gain from its efficient application - but then he/she might be too close to it or have a blinkered view depending upon the project pressures at any given time. On the other hand, would an independent audit team have a sufficiently detailed knowledge of the project to be able to advise confidently on the model results?

7.2.2 Data collection

Creation of a specific cost model will lead to some interesting problems. Common sense or pragmatism may dictate some of the relationships and data types that are required to solve the problem satisfactorily, only to find that during data collection, data needed does not exist, or data exist, but will not be released for use. These situations always occur but nevertheless should be considered useful! If certain data required do not exist, the model can be used to demonstrate the importance or necessity for the company to set about collecting such information because it can be made quite clear what its overall importance is to the final cost estimation (and the company's own control of costs!). After all, 'you cannot control what you do not know'.

If data exist but will not be released for model use, for the usual reasons of sensitivity of data even within the organisation, the same process of sensitivity analysis can demonstrate the importance of access to the data.

The use of somebody else's data is unlikely to be helpful, at least in any detailed way. Obviously, 'foreign' data sets can be used as a broad indicator of basic underlying industry trends and something upon which *some* decisions can be made, but for true indications of the usefulness to your own organisation, your own data must be collected.

Such collection will not be easy. It will need to be collated from many different sources, some of which may not be in your own plant or even company. It has just been mentioned that the data may not exist or if they do may not be released.

It has been estimated that data collection could take several person-months to collect - even longer than basic cost model creation - but is just as, if not more, important.

Maintenance of the data for ongoing use of the model will also be necessary in the same way as needing a computer officer to maintain the software. Such factors do need to be considered when embarking on the route of economics modelling.

7.2.3 Use of economics models

How the model is to be used and applied is critical. With the best will in the world, it is inevitable that errors will affect both the model and the data to which it will be applied. Calculations and formulae will have predictors based upon understanding of practice and estimations of likely outcome. Data input may be incorrect, or based upon figures that are not accurate, i.e. in engineering terms it will be subject to tolerances.

All of these factors will conspire to reduce the basic accuracy of the model and its results despite all efforts to the contrary. In the view of the authors, any use of these models for anything other than providing *additional* information for the decision making process would be a mistake.

Cost model prediction of the final total cost of a project will initially be in error, getting closer to the final *real* cost as the project proceeds to a conclusion.

Such a situation as described above will not necessarily lead to a useful product, for all the hard work put in. The object of the cost modelling that the authors have proposed is to aid in making decisions as the project or design progresses, after which point the decision cannot easily be changed without incurring significant cost increases, which is not what this is all about!

It is advised that such modelling techniques are used merely for *comparing* alternative design/test strategies. In this way any errors in the input data and the calculations will, hopefully, be spread over all results in a broadly similar way.

Economics modelling can be used in many different ways. Obviously there is the application which has been described in this book, i.e. direct application to take part in a detailed design decision making process. In addition to this, benefits are to be found in other applications.

A general spreadsheet model can be made for running on a PC, for example, for management use in creating 'what if' scenarios. This can be in the context of technology choice as well as in test strategy selection. It can, and has, been successfully applied to determining the effects of new ATE investment. But in general, cost modelling can be used for identifying trends and comparing strategies at all levels of the design process - from ASIC design through to system-level design and test, production test and field maintenance.

It can be used in teaching to raise the awareness of test as an important factor in electronic product design. Designers, not forgetting management, are more likely to take notice of what is being said about test if the money involved is mentioned at the same time.

It can be useful even in a research environment to focus attention on those topics that not only require attention, but also will bring a *real* benefit when solved. For example, it can be shown that in some applications efforts to reduce the area overhead associated with a test method would be better applied elsewhere.

Conversely, it can be applied to decide where to start on a large problem when it would not otherwise have been obvious.

Model creation can help those involved in the generation of the model to ensure that they are fully cognisant of the design, manufacturing and test processes involved - model creation can therefore be seen to be a useful educational exercise in its own right.

7.3 RELATED ISSUES

Of direct relevance to the issue of test economics are the areas of system test and field test. The authors have done significant work in these areas but they fall outside the scope of this book. Suffice it to say that they are significant in terms of the overall impact of test costs and *cannot* be ignored. It should also be said that for a true overall analysis of the cost impact of DFT, device testability should not be considered in isolation, unless of course you are working for a merchant chip vendor.

While one form of DFT for a device to be tested in isolation will be appropriate, when it is considered in the context of its application and integration with other components on a board and as part of a system, another form of DFT might be more useful. The total integration of sufficiently complex economics models covering all aspects of system design from chip design to field service and maintenance will be complex, although simplified models may provide useful/usable insights to the scale of the problem.

The models and decision making systems described in this book do not need to be used in isolation but should ideally be integrated into other environments. For example, ECOtest in its current version is a purely advisory tool. The implementation of the test strategy, especially the implementation of the DFT techniques, is not linked to ECOtest. Today's design methodology includes more and more logic synthesis which allows a high level description of the design. A description of DFT techniques on this higher level could be formulated independently of the design. This DFT synthesis description can be part of the test method descriptions of ECOtest, facilitating a link to logic synthesis tools. The DFT part of the generated test strategy is then implemented automatically by a push-button link to the synthesis tool.

The board level test strategy planner (ECOvbs) takes into account a lot of production data, which may be available in a computer integrated manufacturing (CIM) data base. By integrating ECOvbs into such a CIM environment, data could be accessed directly. This reduces the cost of data gathering and makes the provision of data more secure, especially in terms of data updates and data entering errors.

The methods of economics driven test strategy planning, described in this book, are very general and can easily be adopted for other applications of engineering decision making. Test strategy decisions are tightly linked to design and manufacture decisions.

On this basis, other decision making processes, such as technology decisions, or manufacturing strategies, can be integrated into the economics based test strategy planning system. This would extend the test strategy planning system into an engineering decision making system, which would provide a real concurrent engineering tool.

It must be remembered that a system consists of hardware *and* software. This text has dealt exclusively with hardware testing at the chip and board level. However, once the complete system is considered, the importance of software testing becomes evident. Faults occurring during system testing and in the field are often attributed exclusively to the hardware element, when in fact the problem may be due to a software fault. Software testing guidelines should be as rigorous as for hardware. Although software testing is very different than hardware testing, it is just as important and should not be forgotten in the quest for a fault free system.

7.3.1 Time to market

One important aspect in the economic analysis is the time to market. In the economics models described, this was only included in the form of modelling penalty clauses for late delivery. In reality, time-to-market considerations are much more complex. Early introduction of a product results in an increased sales window, as typically the product does not become obsolete any earlier. There is also an associated benefit from gaining a larger market share. DFT often helps minimise the development cycle, easing debugging and reducing expensive redesign. On the other hand, a product introduced hastily with inadequate testing may destroy a company's reputation. Therefore, a trade-off often exists between product quality/functionality and time to market. The situation of course is modified to account for the competition. To complicate things even further, state-of-the-art technologies sometimes have longer development cycles than initially expected. For example, in the case of multi-chip modules (MCMs) it is possible to argue that by the time a successful MCM has been produced (with the associated problems of procuring known good die), a similar result would be achievable using the next generation of VLSI within similar timescales, which can be as long as one or two years. This is obviously not the case in all situations, but it deserves careful thought. The authors have not so far created a comprehensive set of time-to-market models, because they are extremely difficult to generalise. However, this does not diminish the importance of time-to-market considerations, which have a place in any financial analysis.

7.3.2 Qualitative test method comparison

In this text, we have been talking about test costs. Should we need to consider test costs at all? Shouldn't test be an integral part of the design specification in that same way that functionality is? The answer to this question is an emphatic yes. If testability is part of the design specification, then testing can no longer be viewed as an overhead. So why bother with an economic analysis if we are going for full testing anyway? Economic analysis would appear to be a justifiable way to proceed with quantifying different test methods. However, the impact that they each have will vary, and to different extents, i.e. different test methods will not cost the same as each other, but also their relative costs will vary depending upon the application, production volumes, etc. There will be instances where test method A is appropriate, i.e. the cheapest way to achieve required fault coverage, say, but a change in production volume for the same circuit will mean that test method B is now the most cost effective for the same fault cover.

Instead, it has been suggested that test methods should be associated with a 'quality improvement factor' (QUIF) [Bennetts92]. It can be seen from above that a single fixed QUIF value is not always meaningful. Thus the concept of QUIF would seem to be best served by providing each test method with the necessary QUIF formulae that will enable a quick calculation of its financial impact on a given design. The focus of the analysis is no longer on whether testability provision is justifiable, but on which is the most cost effective way of providing the required quality level.

'Quality' is a very emotive word, and can conjure up different responses from different people. But, when applied to electronic products, the term takes on a greater significance - in this context, what is 'quality'?

At a simplistic level, quality can be measured by fault coverage of a test process. Thus the cost of quality can be measured in a similar way to that described previously, i.e. related to factors such as the cost of area overhead needed to achieve high fault coverage. However, such an analysis fails to take into account the wider issues and a strict adherence to mere fault coverage measures may be missing the point by a long way.

'Quality is perception' - what is perceived as being quality by one person will not necessarily be seen as such by another, even for the same product. Consider someone purchasing a television which breaks down, say, after 15 months. If, on reporting the fault to the manufacturer the television is immediately replaced, then quality is perceived. Conversely, if nothing goes wrong with a similar product, the owner will perceive increased quality in the product that was easily replaced [Gomez91].

Should we always strive for quality at all stages of the manufacturing process? 'One company had a process that yielded products with approximately 98% perfection - decided to change process to one with 32% yield because productivity was 20-30 times the yield of the older, stable process' [Gomez91]. Comments likes this need to be taken in context (e.g. have test costs been considered?) but they usefully raise the issue.

We must not forget that market niche will determine the required quality levels. If we accept that quality does cost money to achieve, where is quality appropriate? The manufacturer of a low cost games computer will consider the matter in a totally different way than an air traffic control computer or the electronics controlling the operation of a nuclear reactor.

Much is being said about six-sigma quality processes (equivalent to a yield of 99.99966%). This is positive, as it raises awareness of the necessary issues in the engineering world. It is also easy to take such pronouncements out of context. In [Davis94] an example is presented of an electronics system consisting of printed circuit boards and associated components, the design, manufacture and test of which all conform to six sigma levels of quality. Despite this, it is still possible that only approximately 18% of the final system will be defect free in the particular example given!

Finally, the question of legal liability must again be raised. At the 1993 International Test Conference, a panel session on test economics was held at which Professor Steve Szygenda gave the benefit of his experiences as an expert witness in several product litigation cases. It would appear that the issue of test costs, and whether or not to test, pales into insignificance compared to what will need to be paid out in damages if a product fails through inadequate test or design verification. Even situations that are accepted as being solely attributed to human error (no insinuation of negligence) will end in damages being awarded against the designer/test engineer or employer. 'If you don't do it, you could be sued!', and that will cost you a lot of money.

APPENDIX A

ECOtest economics model description

ECONOMICS MODEL DESCRIPTION FOR DEMO_CIRC

NO DESIGN FOR TEST METHOD SELECTED

nre (Non_recurring_engineering_cost) = 40000.00
vol (Volume) = 5000.00
fun1 (Functions(node1)) = 1.00
fun2 (Functions(node2)) = 3.00
fun3 (Functions(node3)) = 3.00
fun4 (Functions(node4)) = 1.00
fun (Functions) = fun1+fun2+fun3+fun4= 8.00
cells1 (Cells(node1))= 1.00
cells2 (Cells(node2)) = 2000.00
cells3 (Cells(node3)) = 2000.00
cells4 (Cells(node4)) = 1.00
cells (Cells) = cells1+cells2+cells3+cells4= 4002.00
blocks (number of blocks) = 4.00
fdblocks (Functional_described_blocks) = 2.00
kdes (Designer_factor) = 1.30
hpw (Working_hours_per_week) = 40.00
exper (Designers_experience) = 5.00
pcad (Productivity_cad) = 1.00
kp (Productivity_factor) = 1500000.00
in (Inpins) = 33.00
out (Outpins) = 3.00
bi (Bipins) = 16.00
cperf1 (Performance_complexity(node1)) = 1.00
cperf2 (Performance_complexity(node2)) = 1.00
cperf3 (Performance_complexity(node3)) = 1.00
cperf4 (Performance_complexity(node4)) = 1.00
or1 (Originality(node1)) = 1.00
or2 (Originality(node2)) = 0.00
or3 (Originality(node3)) = 0.00
or4 (Originality(node4)) = 1.00
cgate1 (Gates(node1)) = 200.00
cgate2 (Gates(node2)) = 5000.00

cgate3 (Gates(node3)) = 5000.00
cgate4 (Gates(node4)) = 6000.00
cgate (Gates) = cgate1+cgate2+cgate3+cgate4= 16200.00
cexp (Complexity_exponent) = 0.80
percuse (Design_centre_usage) = 0.00
descentrate (Costrate_of_design_centre) = 40000.00
cputime (CPU_time_of_mainframe) = 0.00
equrate (Costrate_of_mainframe) = 6000.00
costrate (Costrate_of_designer) = 4000.00
fpg (Average_fault_per_gate) = 3.00
sdlim (ATPG_99_percent_sequential_limit) = 4.00
glim (ATPG_99_percent_combinational_limit) = 40000.00
tppf (Average_number_of_test_pattern_per_fault) = 0.25
fcreq1 (Required_fault_coverage(node1)) = 0.99
fcreq2 (Required_fault_coverage(node2)) = 0.99
fcreq3 (Required_fault_coverage(node3)) = 0.99
fcreq4 (Required_fault_coverage(node4)) = 0.99
fcv (Typical_fault_coverage_from_verification) = 0.80
tpf1 (Average_time_to_generate_one_test_by_hand(node1)) = 0.10
tpf2 (Average_time_to_generate_one_test_by_hand(node2)) = 0.25
tpf3 (Average_time_to_generate_one_test_by_hand(node3)) = 0.25
tpf4 (Average_time_to_generate_one_test_by_hand(node4)) = 0.05
tpver (Typical_number_of_test_patterns_from_simulation) = 10000.00
pms (Test_vector_limit) = 32000.00
pps (Additional_price_per_step) = 5.00
avs1 (Average_sequential_depth(node1)) = 0.00
avs2 (Average_sequential_depth(node2)) = 9.00
avs3 (Average_sequential_depth(node3)) = 9.00
avs4 (Average_sequential_depth(node4)) = 3.00
ktgs (Factor_for_ATPG) = 0.0000004
exptgs (Complexity_exponent_for_ATPG) = 2.00
ppu (Price_per_unit) = 0.00
npu (Non_delivered_units)= 0.00
npcost (Penalty_for_non_performance) = 0.00
puc (Production_unit_cost) = 16.20
pc (Production_cost) = nre+vol*puc= 120995.59
plib (Productivity_of_library) =
kp*(fun/(cells+blocks)+2*fun/(cells+fdblocks))/3= 2996.50
pdes (Experience_of_designer) = 1-1/(kdes+exper)= 0.84
prod (Productivity) = pcad*pdes*plib= 2520.87
cpin (Pin_complexity) = 2*in+out+3*bi= 117.00
orabs1 (Absolute_originality(node1)) = (1-or1)*cgate1= 0.00
orabs2 (Absolute_originality(node2)) = (1-or2)*cgate2= 5000.00

orabs3 (Absolute_originality(node3)) = (1-or3)*cgate3= 5000.00
orabs4 (Absolute_originality(node4)) = (1-or4)*cgate4= 0.00
orabs (Absolute_originality) = orabs1+orabs2+orabs3+orabs4= 10000.00
or (Originality) = orabs/cgate= 0.62
cperfabs1 (Absolute_performance(node1)) = cperf1*cgate1= 200.00
cperfabs2 (Absolute_performance(node2)) = cperf2*cgate2= 5000.00
cperfabs3 (Absolute_performance(node3)) = cperf3*cgate3= 5000.00
cperfabs4 (Absolute_performance(node4)) = cperf4*cgate4= 6000.00
cperfabs (Absolute_performance = cperfabs1+cperfabs2+cperfabs3+cperfabs4= 16200.00
cperf (Performance_complexity) = cperfabs/cgate= 1.00
lcompl1 (Local_complexity(node1)) = cperf1*(1-or1)*cgate1= 0.00
lcompl2 (Local_complexity(node2)) = cperf2*(1-or2)*cgate2= 5000.00
lcompl3(Local_complexity(node3)) = cperf3*(1-or3)*cgate3= 5000.00
lcompl4 (Local_complexity(node4)) = cperf4*(1-or4)*cgate4= 0.00
lcompl (Local_complexity) = lcompl1+lcompl2+lcompl3+lcompl4= 10000.00
ocompl (Overall_complexity) = cpin*lcompl^cexp= 185432.52
mp (Manpower) = ocompl/prod= 73.56
destime (Design_time) =
(9.2*mp^0.34>29)*(9.2*mp^0.34)+(9.2*mp^0.34<29)*mp+(9.2*mp^0.34=29)*mp= 39.67
engcost (Engineering_cost) = mp*costrate= 294235.91
descentcost (Design_centre_cost) = percuse*destime*descentrate= 0.00
mainframecost (Mainframe_cost) = cputime*equrate= 0.00
descost (Design_cost) = engcost+descentcost+mainframecost= 294235.91
faults1 (Number_of_faults(node1)) = cgate1*fpg= 600.00
faults2 (Number_of_faults(node2)) = cgate2*fpg= 15000.00
faults3 (Number_of_faults(node3)) = cgate3*fpg= 15000.00
faults4 (Number_of_faults(node4)) = cgate4*fpg= 18000.00
faults (Number_of_faults = faults1+faults2+faults3+faults4= 48600.00
fach1 (Achievable_fault_cover(node1)) =
(1-2.72^(-4.6*sdlim/avs1))*(1-2.72^(-4.6*glim/cgate1))= 1.00
fach2 (Achievable_fault_cover(node2)) =
(1-2.72^(-4.6*sdlim/avs2))*(1-2.72^(-4.6*glim/cgate2))= 0.87
fach3 (Achievable_fault_cover(node3)) =
(1-2.72^(-4.6*sdlim/avs3))*(1-2.72^(-4.6*glim/cgate3))= 0.87
fach4 (Achievable_fault_cover(node4)) =
(1-2.72^(-4.6*sdlim/avs4))*(1-2.72^(-4.6*glim/cgate4))= 1.00
remfaults1 (Remaining_faults_for_MTPG(node1)) =
((fcreq1-fcach1)>0)*((fcreq1-fcach1)*faults1)= 0.00
remfaults2 (Remaining_faults_for_MTPG(node2)) =
((fcreq2-fcach2)>0)*((fcreq2-fcach2)*faults2)= 1789.27

remfaults3 (Remaining_faults_for_MTPG(node3)) =
((fcreq3-fcach3)>0)*((fcreq3-fcach3)*faults3)= 1789.27
remfaults4 (Remaining_faults_for_MTPG(node4)) =
((fcreq4-fcach4)>0)*((fcreq4-fcach4)*faults4)= 0.00
mtgtime1 (Time_for_MTPG(node1)) = remfaults1*tpf1/hpw= 0.00
mtgtime2 (Time_for_MTPG(node2)) = remfaults2*tpf2/hpw= 11.18
mtgtime3 (Time_for_MTPG(node3)) = remfaults3*tpf3/hpw= 11.18
mtgtime4 (Time_for_MTPG(node4)) = remfaults4*tpf4/hpw= 0.00
mtgtime (Time_for_MTPG) = mtgtime1+mtgtime2+mtgtime3+mtgtime4=
22.37
mtc (Cost_of_MTPG) = mtgtime*costrate= 89463.73
faultsa1 (Number_of_faults_for_ATPG(node1)) =
((fcach1-fcv)>0)*((fcach1-fcv)*faults1)= 120.00
faultsa2 (Number_of_faults_for_ATPG(node2)) =
((fcach2-fcv)>0)*((fcach2-fcv)*faults2)= 1060.73
faultsa3 (Number_of_faults_for_ATPG(node3)) =
((fcach3-fcv)>0)*((fcach3-fcv)*faults3)= 1060.73
faultsa4 (Number_of_faults_for_ATPG(node4)) =
((fcach4-fcv)>0)*((fcach4-fcv)*faults4)= 3600.00
numtpa1 (Number_of_ATPG_pattern(node1)) = faultsa1*tppf*(avs1+1)=
16777216.00 (initialised separately to exhaustive test set)
numtpa2 (Number_of_ATPG_pattern(node2)) = faultsa2*tppf*(avs2+1)=
2651.81
numtpa3 (Number_of_ATPG_pattern(node3)) = faultsa3*tppf*(avs3+1)=
2651.81
numtpa4 (Number_of_ATPG_pattern(node4)) = faultsa4*tppf*(avs4+1)=
33024.00
numtpm1 (Number_of_MTPG_pattern(node1)) = remfaults1*(avs1+1)= 0.00
numtpm2 (Number_of_MTPG_pattern(node2)) = remfaults2*(avs2+1)=
17892.75
numtpm3 (Number_of_MTPG_pattern(node3)) = remfaults3*(avs3+1)=
17892.75
numtpm4 (Number_of_MTPG_pattern(node4)) = remfaults4*(avs4+1)= 0.00
tpgen1 (Number_of_generated_test_pattern(node1)) = numtpa1+numtpm1=
16777216.00
tpgen2 (Number_of_generated_test_pattern(node2)) = numtpa2+numtpm2=
20544.56
tpgen3 (Number_of_generated_test_pattern(node3)) = numtpa3+numtpm3=
20544.56
tpgen4 (Number_of_generated_test_pattern(node4)) = numtpa4+numtpm4=
33024.00
sumtp (Number_of_MTPG_plus_ATPG_pattern) = 16830784 (exhaustive PLA test)
tsl (Test_set_length) = sumtp+tpver= 16840784.00

tac (Test_application_cost) = (tsl-tsl:pms)/pms*pps*vol= 13150000.00
costtgs1 (Cost_for_test_pattern_generation(node1)) =
ktgs*(cgate1*(avs1+1))^exptgs= 0.02
costtgs2 (Cost_for_test_pattern_generation(node2)) =
ktgs*(cgate2*(avs2+1))^exptgs= 1000.00
costtgs3 (Cost_for_test_pattern_generation(node3)) =
ktgs*(cgate3*(avs3+1))^exptgs= 1000.00
costtgs4 (Cost_for_test_pattern_generation(node4)) =
ktgs*(cgate4*(avs4+1))^exptgs= 230.40
costtgs (Cost_for_test_pattern_generation) =
costtgs1+costtgs2+costtgs3+costtgs4= 2230.42
tcost (Test_cost = tac+mtc+costtgs)= 13241694.00
plcost (Cost_for_profit_loss) = ppu*ndu= 0.00
ttmcost (Time_to_market_cost) = npcost+plcost= 0.00
ovcost (Overall_cost) = pc+descost+tcost+ttmcost= 13656926.00

APPENDIX B

Design description of demo_circ

DEMO_CIRC DESIGN DATA

Name of blocktype : demo_circ
Level : 0
List of subblocks : (4) node1(block1), node2(block2), node3(block2), node4(block3)
List of bundles : (13) s_1, *cl*, s_3, s_4, s_5, s_6, s_7, s_8, s_9, s_{10}, s_{11}, s_{12}, *bus*

Netlist-independent data

Number of equivalent gates	: 16200
Number of equivalent D-FFs	: 4646
Number of cells	: 4002
Time for MTPG per fault (h)	: 0.80
Average sequential depth	: 25
Design style	: Unknown
Design class	: Sequential random logic
Originality	: 0.38
FD-Blocks	: 2
Number of functions	: 8
Gate limit	: 28000
Pin limit	: 64
Max. self test time (s)	: 60

Transparency functions
(none)

Name of blocktype : block1
Level : 1
List of subblocks : (0)
List of bundles : (4) i_1, i_2, o_1, o_2

Netlist-independent data

Number of equivalent gates	: 200
Number of product terms	: 50
Number of cells	: 1

Time for MTPG per fault (h) : 0.10
Average Sequential depth : 0
Design style : Asynchronous design
Design Class : PLA
Originality : 1.00
FD-Blocks : 1
Number of functions to perform : 1
Performance complexity factor : 1.00
Required fault cover : 99.0%

Transparency functions

Transparency from i_1 to o_2 is i/o transparent

Name of blocktype : block2
Level : 1
List of subblocks : (0)
List of bundles : (6)cl, i_2, i_3, b, o_1, o_2

Netlist-independent data

Number of equivalent gates : 5000
Number of equivalent D-FFs : 250
Number of cells : 2000
Time for MTPG per fault (h) : 0.25
Average Sequential depth : 9
Design style : Flip-flop design
Design Class : Sequential random logic
Originality : 0.00
FD-Blocks : 0
Number of functions to perform : 3
Performance complexity factor : 1.00
Required fault cover : 99.0%

Transparency Functions

(none)

Name of blocktype : block3
Level : 1
List of subblocks : (0)
List of bundles : (7) i_1, cl, i_3, rw, b, o_1, o_2

Netlist-independent data

Number of equivalent gates : 6000

Memory size (in bits)	: 4096
Number of cells	: 1
Time for MTPG per fault (h)	: 0.05
Average sequential depth	: 3
Design style	: Flip-flop design
Design class	: RAM
Originality	: 1.00
FD-Blocks	: 1
Number of functions to perform	: 1
Performance complexity factor	: 1.00
Required fault cover	: 99.0%

Transparency functions
Transparency from i_3 to o_1 is Control-line transparent
Transparency from rw to o_2 is i/o transparent

APPENDIX C

An example board level economics model description

EVEREST-VBS COST MODEL

CURRENCY: DM

NAME OF PBA TYPE: PB1

1. BASIC ASSUMPTIONS

Application period	T=0	
Hours per working week	HPW	37
Hours per working month	HPM	155
Annual working hours	AWH	1860
Market interest rate/ year	Mi	12%
Depreciation period of ATE in years (as systems production time)	Depy	6

		STRATEGY 1 (without boundary scan)	STRATEGY 2 (with boundary scan)
Number of boards to be produced	tnpba1	5047	5049
2. DEVELOPEMENT PHASE			
Iterations before production	Ipre	1	1
Iterations after production	Ipost	0.5	0.5
2.1 Specification			
Initial spec. time (h)	Tspec	400	400
Hourly labour rate	Kspec	DM120	DM120
Specification labour cost	Cspec	"DM48,000"	"DM48,000"
2.2 Design (Entry + Simulation)			
Aver.time taken for initial design	Tid	600	620
Design iteration factor	d	20%	20%
Hourly design labour rate	Kde	DM120	DM120

Hourly CAE softw.& equipment rate	KCAE	DM0	DM0
Time taken to complete design (h)	Tdes	737	761
Total labour cost for design	Cdes	"DM88,390"	"DM91,336"
Total cost of CAE softw.& eq.	Cdeq	DM0	DM0
Total design cost		"DM88,390"	"DM91,336"
2.3 Layout			
Aver.time taken for initial layout	Tilay	150	150
Layout iteration factor	l	10%	10%
Hourly layout labour rate	Kl	DM100	DM100
Hourly layout software & equipment rate	KLAY	DM0	DM0
Time taken to complete layout (h)	Tlay	166.1396204	166.1396204
Total labour cost for layout	Clay	"DM16,614"	"DM16,614"
Total cost of layout s.& eq./h	Cleq	DM0	DM0
Total layout cost		"DM16,614"	"DM16,614"
2.4 Prototype Fabrication			
In-house prototype constr.time (h)	Tip	120	120
Prototype iteration factor	p	90%	90%
Hourly prototype labour rate	Kp	DM100	DM100
Mat.cost for protot.per iteration	Cmat	"DM5,000"	"DM5,700"
Number of prototypes per iteration	Nprot	1	1
Total prototype time (h)	Tp	278	278
Prototype labour cost	Cpl	"DM27,788"	"DM27,788"
Prototype material cost	Cp	"DM12,500"	"DM14,250"
Total prototype cost		"DM40,288"	"DM42,038"
2.5 Verification (prototype test)			
Functional ver. time (h)	Tfv	500	500
Verif. time in the system (h)	Tsysv	500	500
Verification iteration factor	v	80%	80%
Hourly verification labour rate	Kv	DM120	DM120
Hourly verification equipment rate	KVER	DM0	DM0
Total verification time (h)	Tv	2138	2138

Labour verification cost	Cv	"DM256,540"	"DM256,540"
Total cost of verification eq.	Ceqv	DM0	DM0
Total verification cost		"DM256,540"	"DM256,540"

2.6 Test engineering

		STRATEGY 1 (without boundary scan)			STRATEGY 2 (with boundary scan)		
Test stage name		Vi	Ict	Func.	Vi	B/Scan	Funct.+B/S
Test plan	Tpl	0	1	0	0	1	0
Length of test prog. (1000s lines)	Prg	0	500	0	0	550	18
Fraction of test prog. generated automatically	Fga	0%	80%	10%	0%	0%	10%
Length of tool for autom.test.gen.	Tprg	0	0	0	0	0	0
Iteration factor	ift	0%	20%	0%	0%	20%	0%
Hourly test eng. Labour rate	Kte	DM0	DM100	DM120	DM0	DM90	DM120
Hourly test eng. Equipment rate	Ktee	DM0	DM0	DM0	DM0	DM0	DM0
Preparation cost /ATE (fixture)	PrepC	DM0	DM30K	DM0	DM0	DM5K	DM0
Software eng. Time for tools (h)	Tto	0	0	0	0	0	0
Test eng. time for test prog. (h)	Tte	0	259	0	0	296	0

		Strategy 1	Strategy 2
Labour cost for test engineering	Cte	"DM25,900"	"DM26,640"
Total cost for test eng.equip.	Cteq	DM0	DM0
Total test engineering cost		"DM25,900"	"DM26,640"

3. PRODUCTION PHASE

STRATEGY 1 (without boundary scan)	STRATEGY 2 (with boundary scan)

3.1 Manufacture phase

Production prepare Cost (Fixed)	Cpp	"DM4,000"	"DM4,000"
Bare board cost	Cbb	DM100	DM100
Component Cost / board	Cc	"DM5,000"	"DM5,700"
Number of components:			
Axial	nax	20	20
Radial	nrad	5	5
ICs	nic	20	20
SMDs	nsmd	35	35
Manual	nman	20	20
Others	noth	0	0
Assembly cost:			
Axial	Caax	DM0.10	DM0.10
Radial	Carad	DM0.10	DM0.10
ICs	Caic	DM0.15	DM0.15
SMDs	Casmd	DM0.10	DM0.10
Manual	Caman	DM0.40	DM0.40
Others	Caoth	DM0.10	DM0.10
Overall assembly Cost / Board	Cass	DM17	DM17
Production variable Cost/per board	Bvar	"DM5,117"	"DM5,817"
Total production Var. cost	Pvar	"DM25,823,933"	"DM29,372,040"
Tot. production Prep.(fixed) cost	Pfix	"DM4,000"	"DM4,000"
Total production cost (For TNPA)	TCNPB	"DM25,827,933"	"DM29,376,040"
Production unit cost for boards going to systems assembly	PUSA	"DM5,166"	"DM5,875"

3.2 Test phase

Number of shifts	NS	1	
Max. prod. Vol. / year	PV	1500	
Failure rates:			
Shorts per 100 PBA	Fsh	7	7.1
Opens per 100 PBA	Fop	8	8.1
Wrong/	Fpla	4	4

missing compon. per 100 PBA			
Faulty an.& dig.ICs	Fan	10	10.1
Dynamic faults	Fdig	1	1
Other faults per 100 PBA	Foth	6.5	6.6
Total faults /100 boards	FPB	37	37
Production yield	Yp	69%	69%
Max. MTBF in early life phase (h)	MTBF1	1000	1000
MTBF after early life (h)	MTBF2	6000	6000
Early life time (h)	telf	200	200

		STRATEGY 1 (without boundary scan)			STRATEGY 2 (with boundary scan)		
Test stage name		Vi	Ict	Func.	Vi	B/Scan	Funct.+B/S
3.2.1 Quality model							
Number of iterations in test	nit	3	3	3	3	3	3
Shorts fault coverage	FCsh	5%	90%	100%	5%	95%	100%
Opens fault coverage	FCop	5%	90%	75%	5%	95%	75%
Wrong/missing comp. fault coverage	FCpla	80%	100%	50%	80%	100%	50%
Fault cov. of an.& dig.	FCan	0%	90%	65%	0%	90%	65%
Dynamic fault coverage	FCdig	0%	0%	0%	0%	0%	0%
Other	FCoth	10%	80%	100%	10%	80%	100%
Shorts going to test	USh	7.0	7.0	0.7	7.1	7.1	0.4
Opens going to test	Uop	8.0	8.0	0.8	8.1	8.1	0.4
Wrong/miss. comp. going to test	Uwr	4.0	4.0	0.0	4.0	4.0	0.0
Faulty an.& dig. w/o BS gtT.	Uan	10.0	10.0	1.0	10.1	10.1	1.0
Faulty digit. going to test	UnDig	1.0	1.0	1.0	1.0	1.0	1.0
Other faults going to test	UnOth	6.5	6.5	1.3	6.6	6.6	1.3
Early life failures	elf	0.0	0.0	0.0	0.0	0.0	0.0
Total Faults going to test/100 PBA	Tfun	36.5	36.5	4.8	36.9	36.9	4.1
Detected Shorts	DetSh	0.00	6.30	0.00	0.00	6.75	0.00

Detected Opens	DetOp	0.00	7.20	0.00	0.00	7.70	0.00
Detected wrong/missing components	DetPla	0.00	4.00	0.00	0.00	4.00	0.00
Detected defective an.& dig. w/o BS	DetAn	0.00	9.00	0.00	0.00	9.09	0.00
Detected defective BS components	DetDig	0.00	0.00	0.00	0.00	0.00	0.00
Detected other	DetOth	0.00	5.20	0.00	0.00	5.28	0.00
Detected early life failures	DetElf	0.00	0.01	0.00	0.00	0.01	0.00
Total detected faults per 100 PBAs	ds	0.0	31.7	0.0	0.0	32.8	0.0
Yield after test	Yat	69%	95%	95%	69%	96%	96%
Test iteration factor	Tit	21%	29%	4%	21%	30%	4%
Max. number of boards going to test /Y	Ntest	1900	2102	1570	1903	2115	1560
Max.num.of boards going to diag./rep./Y	Ndr	400	602	70	403	615	60
Number of non-repairable boards /Y	Nnr	3	11	0	3	12	0
Total number of boards going to test	TNtest	6394	7071	5283	6406	7119	5251
Total Num.of boards going to diag./rep.	TNdr	1347	2025	236	1356	2070	202
Total Number of non-rep. boards	TNnr	10	36	0	10	39	0

3.2.2 Test time model

Lot size	Lot	1	25	1	1	25	1
Acceleration factor (accel.life test)	A	1	1	0	1	1	1
Set up time/lot (minutes)	Tsu	0	20	0	0	20	20
Attended test time for good boards (min)	Tatt	0	5	0	0	5	0
Unattended test time (minutes)	Tun	0	1	0	0	1	0
Diagnosis time per fault (minutes)	Tdiag	0	2	0	0	2	0
Repair time per fault (minutes)	Trep	0	20	0	0	20	0
			STRATEGY 1			STRATEGY 2	
Accelerated test time	ta	0	6	0	0	6	0

Accumulated test time	tacc	0	6	6	0	6	6
Average MTBF for early life fail.	amtbf	1000	1001.25	1002.5	1000	1001.25	1002.5
Effective test time for Tteff good board (minutes)	0.00	6.80	0.00	0.00	6.80	20.00	
diagnosis & repair time / board	Td	0.00	22.00	0.00	0.00	22.00	0.00
Total test time/yr. (hours)	TTY	0	238	0	0	240	0
Total diag/rep time per year (h)	Tdryr	0	221	0	0	225	0
Testers required	TER	0.00	1.00	0.00	0.00	1.00	0.00
Utilisation grade	util	0%	24.7%	0%	0%	25%	0%

3.2.3 Test cost model

Hourly test labour rate	Kt	DM0	DM75	DM0	DM0	DM75	DM0
Hourly var. operating Cost per ATE	KATE	DM0	DM0	DM0	DM0	DM0	DM0
Labour cost of testing / board	Ct	DM0	DM7	DM0	DM0	DM7	DM0
Labour cost of diagnosis / board	Cd	DM0	DM28	DM0	DM0	DM28	DM0
"Test,diagnosis, repair labour costs"	PRC	DM0	"DM 106,946"	DM0	DM0	"DM 108,537"	DM0
Tot. preparation cost per test stage	SpreC	DM0	"DM 30K"	DM0	DM0	"DM 5K"	DM0
Purchase cost for ATE	PATE	DM0	"DM 500,000"	DM0	DM0	"DM 60,000"	DM0
Fixed operating costs /Y	OfATE	DM0	"DM 50,000"	DM0	DM0	"DM 6,000"	DM0
Depreciation cost /Y for ATE	DcATE	DM0	"DM 83,333"	DM0	DM0	"DM 10,000"	DM0
interest cost /Y for ATE	IcATE	DM0	"DM 30,000"	DM0	DM0	"DM 3,600"	DM0
Total LC fixed costs	TATE	DM0	"DM	DM0	DM0	"DM	DM0

for 1 ATE (During systems production time)			980,000"			117,600"	
Equipment fixed cost per test stage	Cpeq	DM0	"DM 980,000"	DM0	DM0	"DM 117,600"	DM0
Var. oper. costs for ATE's /board	OvATE	DM0	DM0	DM0	DM0	DM0	DM0
ATE Var.oper.costs	OOVC		DM0	DM0	DM0	DM0	DM0	DM0

		STRATEGY 1 (without boundary scan)	STRATEGY 2 (with boundary scan)
Overall labour T & R cost	APTR	"DM106,946"	"DM108,537"
Total LC var. operating costs	TOv	DM0	DM0
"Total fixed test,d.,rep. eq. cost "	Cpte	"DM980,000"	"DM117,600"
Total test preparation cost	TpreC	"DM30,000"	"DM5,000"
Test-related labour cost per board		DM21.39	DM21.71
Total variable operating costs per board		DM0.00	DM0.00
"Total fixed test,d.r. eq. cost per board"		DM196.00	DM23.52
Cost of preparation per board	DM6.00	DM1.00	
Remaining faults/100 PBAs:			
Shorts	remsh	0.70	0.36
Opens	remop	0.80	0.41
Wrong/missing components	rempla	0.00	0.00
Defective an./dig. comp. w/o BS	reman	1.00	1.01
Defective BS components	remdig	1.00	1.00
Others	remoth	1.30	1.32
Total remaining faults / 100 PBA:	remsum	4.80	4.09
Remaining faults per board	frem	4.80E-02	4.09E-02
Overall board test time (min)	btt	6	6

4. COST SUMMARIES FOR PB1

4.1 Developement costs

Labour	dlab	"DM463,232"	"DM466,918"
Material	dmat	"DM12,500"	"DM14,250"
Equipment	dequ	DM0	DM0
Total developement costs	NRE	"DM475,732"	"DM481,168"

Developement beginning year	DBY	0	0
Developement period in years	DP	1	1
Total developement costs at T=0	NRE0	"DM424,761"	"DM429,615"

4.2 Production costs

4.2.1 Manufacturing costs

Tot. production prep.(fixed) cost	Pfi	"DM4,000"	"DM4,000"
Tot. production var. cost	Pva	"DM 25,823,933"	"DM29,372,040"
Manufacturing beginning year	MBY	1	1
Manufacturing period in years	MP	5	5
Tot. production fixed cost at T=0	Pfix0	"DM3,571"	"DM3,571"
Tot. prod. var. cost at T=0	Pvar0	"DM 16,623,125"	"DM18,907,077"
Tot. prod. fixed cost per board	Pfix	DM0.71	DM0.71
Tot. prod. var. cost per board	Pvar	"DM3,325"	"DM3,781"
Total production cost at T=0	tpc	"DM 16,626,696"	"DM18,910,648"

4.2.2 Test cost

Test-related labour cost	tla	"DM106,946"	"DM108,537"
Variable operating costs	tva	DM0	DM0
"Fixed test,d/r. eq. cost "	tfi	"DM980,000"	"DM117,600"
Test equ. preparation cost	tpre	"DM30,000"	"DM5,000"
Total test cost	tbtc	"DM 1,116,946"	"DM231,137"
Test beginning year	TBY	1	1
Test period in years	TPY	5	5
Test-related labour cost at T=0	tlab0	"DM68,842"	"DM69,867"
Variable operating costs at T=0	tvar0	DM0	DM0
"Fixed test,d/r. eq. cost at T=0	tfix0	"DM875,000"	"DM105,000"
Test equ. preparation cost at T=0	tprep0	"DM19,311"	"DM3,219"
Test-related labour cost /board	tlab	DM13.77	DM13.97
Variable operating costs /board	tvar	DM0.00	DM0.00
"Fixed test,d/r. eq. cost /board"	tfix	DM175.00	DM21.00
Test equ. preparation cost /board	tprep	DM3.86	DM0.64
Total test cost at T=0 ttc	"DM963,154"	"DM178,085"	
Total number of boards going to system assembly	TNSA	5000	5000

TOTAL BOARD COST	OV	"DM 18,014,611"	"DM19,518,348"
TOTAL COST PER BOARD	LccpB	"DM3,603"	"DM3,904"

Appendix D

Test method descriptions used in the ECOtest system

Test method descriptions given here consist of an information file, which is a text description for the benefit of the designer, and a database file, used by the system.

NO APPLICATION OF DESIGN FOR TESTABILITY

DESCRIPTION:
This test method assumes that no special DFT method will be applied to increase the testability of the block. Ext_nodft assumes that no special DFT method will be applied to increase the accessibility of the block.

TEST VECTORS:
The test pattern generation method depends on the sequential depth of the block. For combinational blocks, an ATPG system for combinational and full scan circuits can be used. For sequential circuits, the applicability of an ATPG system depends on the sequential depth of the circuit, and whether the block is a synchronous or asynchronous design.

TEST VECTORS:
no impact.

EXTRA GATES:
No extra gates.

EXTRA PINS:
No extra pins.

PERFORMANCE IMPACT:
none.

REFERENCE:
none.

```
NAME int_nodft
TYPE intern
DESIGN_CLASS RS
DESIGN_CLASS RC
DESIGN_CLASS MU
```

```
NAME ext_nodft
TYPE extern
DESIGN_CLASS RS
DESIGN_CLASS RC
DESIGN_CLASS PL
```

DESIGN_CLASS AL
TPG_METHOD N
SELF_TEST n
SEQ_DEPTH $SDEPTH
OV_G 0
ACCESS_IN_D n
ACCESS_IN_CO n
ACCESS_IN_CL n
ACCESS_OUT n
ACCESS_BUS n
PERF 1
OR 1
OV_INPIN 0
OV_OUTPIN 0
OV_BIPIN 0
PIN_COMP 0
D_STYLE A
D_STYLE F
D_STYLE L
FUN 0
ACH_FC C
NUM_TP C
CELLS 0

DESIGN_CLASS RA
DESIGN_CLASS RO
DESIGN_CLASS MU
DESIGN_CLASS AL
TPG_METHOD N
SELF_TEST n
SEQ_DEPTH $SDEPTH
OV_G 0
ACCESS_IN_D n
ACCESS_IN_CO n
ACCESS_IN_CL n
ACCESS_OUT n
ACCESS_BUS n
PERF 1
OR 1
OV_INPIN 0
OV_OUTPIN 0
OV_BIPIN 0
PIN_COMP 0
D_STYLE A
D_STYLE F
D_STYLE L
FUN 0
ACH_FC C
NUM_TP C
CELLS 0

INT_SCAN: Internal scan path

DESCRIPTION:
Every flip-flop is replaced by a scan path flip-flop. All flip-flops are connected to a scan register. The block can be modelled as combinational logic for testing purposes.

TEST VECTORS:
dependent on logic function

EXTRA GATES:
number_of_flip-flops*2.5

EXTRA PINS:
2 inputs
1 output

PERFORMANCE IMPACT:

2 gate delays per pin.

REFERENCE:
[Williams83

NAME int_scan
TYPE intern
DESIGN_CLASS RS
TPG_METHOD C
SELF_TEST n
SEQ_DEPTH 0
OV_G $DFF*2.5
ACCESS_IN_D n
ACCESS_IN_CO n
ACCESS_IN_CL n
ACCESS_OUT n
ACCESS_BUS n
PERF 1
OR 1
OV_INPIN 2
OV_OUTPIN 1
OV_BIPIN 0
PIN_COMP 1
D_STYLE F
FUN 0
ACH_FC C
NUM_TP C
CELLS 0

L2_LATCH: Internal L2* Latch

DESCRIPTION:
Two single latches are replaced by a shift-register-latch (SRL). All latches are connected to a scan register. The block can be modelled as combinational logic for testing purposes.

TEST VECTORS:
Dependent on logic function.

EXTRA GATES:
number_of_latches*2

EXTRA PINS:
2 inputs

1 output

PERFORMANCE IMPACT:
1 gate delay per pin.

REFERENCE:
[Ohletz87]

NAME l2_latch
TYPE intern
DESIGN_CLASS RS
TPG_METHOD C
SELF_TEST n
SEQ_DEPTH 0
OV_G $DFF*2
ACCESS_IN_D n
ACCESS_IN_CO n
ACCESS_IN_CL n
ACCESS_OUT n
ACCESS_BUS n
PERF 1.1
OR 0.9
OV_INPIN 3
OV_OUTPIN 1
OV_BIPIN 0
PIN_COMP 2
D_STYLE L
FUN 0
NUM_TP C
ACH_FC C
CELLS 0

S_LATCH: Internal Single Latch

DESCRIPTION:
Every single latches is replaced by a shift register latch (SRL). All latches are connected to a scan register. The block can be modelled as combinational logic for testing purposes.

TEST VECTORS:
Dependent on logic function.

EXTRA GATES:
number_of_latches*5.5

EXTRA PINS:
3 inputs
1 output

PERFORMANCE IMPACT:
1 gate delay per pin.

REFERENCE:
[Ohletz87]

NAME s_latch
TYPE intern
DESIGN_CLASS RS
TPG_METHOD C
SELF_TEST n
SEQ_DEPTH 0
OV_G $DFF*5.5
ACCESS_IN_D n
ACCESS_IN_CO n
ACCESS_IN_CL n
ACCESS_OUT n
ACCESS_BUS n
PERF 1
OR 0.9
OV_INPIN 3
OV_OUTPIN 1
OV_BIPIN 0
PIN_COMP 2
D_STYLE L
FUN 0
NUM_TP C
ACH_FC C
CELLS 0

INT_D_LATCH: Internal Double Latch

DESCRIPTION:
Every double latch is replaced by a shift register latch (SRL). All latches are connected to a scan register. The block can be modelled as combinational logic for testing purposes.

TEST VECTORS:
Dependent on logic function.

EXTRA GATES:
number_of_latches

EXTRA PINS:
2 inputs
1 output

PERFORMANCE IMPACT:
1 gate delay per pin.

REFERENCE:
[Ohletz87

NAME int_d_latch
TYPE intern
DESIGN_CLASS RS
TPG_METHOD C
SELF_TEST n
SEQ_DEPTH 0
OV_G $DFF
ACCESS_IN_D n
ACCESS_IN_CO n
ACCESS_IN_CL n
ACCESS_OUT n
ACCESS_BUS n
PERF 1
OR 0.9
OV_INPIN 3
OV_OUTPIN 1
OV_BIPIN 0
PIN_COMP 2
D_STYLE L
FUN 0
NUM_TP C
ACH_FC C
CELLS 0

INT_SCAN_SET: Internal Scan-Set

DESCRIPTION:
To every storage element a scan-set flip-flop is added. All flip-flops are connected to a scan register. The block can be modelled as combinational logic for testing purposes.

TEST VECTORS:
Dependent on logic function.

EXTRA GATES:
number_of_flip-flops*18

EXTRA PINS:
3 inputs
1 output

PERFORMANCE IMPACT:
2 gate delays per pin.

REFERENCE:
[Williams83]

NAME int_scan_set
TYPE intern
DESIGN_CLASS RS
TPG_METHOD C
SELF_TEST n
SEQ_DEPTH 0
OV_G $DFF*18
ACCESS_IN_D n
ACCESS_IN_CO n
ACCESS_IN_CL n
ACCESS_OUT n
ACCESS_BUS n
PERF 1.1
OR 0.95
OV_INPIN 3
OV_OUTPIN 1
OV_BIPIN 0
PIN_COMP 1
D_STYLE A
D_STYLE F
D_STYLE L
FUN 0
NUM_TP C
ACH_FC C
CELLS 0

BILBO: Built-In Logic Block Observer

DESCRIPTION:
A special architecture of a linear feedback shift register for the usage of self-test.

TEST VECTORS:
(2^inputs) - 1

EXTRA GATES:
(outputs+inputs)*12+3

EXTRA PINS:
3 inputs
1 output

REFERENCE:
[Koenemann79

NAME bilbo
TYPE intern
DESIGN_CLASS RC
TPG_METHOD E
SELF_TEST y
SEQ_DEPTH $SDEPTH
OV_G ($OUTPINS+$INPINS)*12+3
ACCESS_IN_D y
ACCESS_IN_CO y
ACCESS_IN_CL n
ACCESS_OUT y
ACCESS_BUS n
PERF 1.12
OR 0.9
OV_INPIN 3
OV_OUTPIN 1
OV_BIPIN 0
PIN_COMP 10
D_STYLE A
D_STYLE F
D_STYLE L
FUN 0
NUM_TP 2^$INPINS-1
ACH_FC 1
CELLS 0

BILBO_SCAN: Built-In Logic Block Observer

DESCRIPTION:
A special architecture of a linear feedback shift register for the usage of self test. In addition, every flip-flop must be replaced by a scan-path flip-flop.

TEST VECTORS:
(2^(inputs+number_of_flip_flops)) - 1

EXTRA GATES:
(outputs+inputs)*12+3 + 2.5*number_of_flip_flops

EXTRA PINS:
3 inputs
1 output

REFERENCE:
[Koenemann79

```
NAME bilbo_scan
TYPE intern
DESIGN_CLASS RS
TPG_METHOD E
SELF_TEST y
SEQ_DEPTH 0
OV_G ($OUTPINS+$INPINS)*12+3+$DFF*2.5
ACCESS_IN_D y
ACCESS_IN_CO y
ACCESS_IN_CL n
ACCESS_OUT y
ACCESS_BUS n
PERF 1.12
OR 0.9
OV_INPIN 3
OV_OUTPIN 1
OV_BIPIN 0
PIN_COMP 10
D_STYLE F
FUN 0
NUM_TP 2^($INPINS+$DFF)-1
ACH_FC 1
CELLS 0
```

CIRCSTP: Circular Self Test Path

DESCRIPTION:
Circular self-test path is an off-line BIST technique which provides both pseudo-random test generation and test response compaction and is based on the feedback shift register approach. System registers are selectively replaced by special BIST registers which are connected in a circular path forming a feedback shift register.

TEST VECTORS:
~ 8*2^inputs

EXTRA GATES:
(number_of_flip_flops-outputs)*4-inputs*1.5

EXTRA PINS:
3 inputs
1 output

REFERENCE:
[Krasniewski88]

```
NAME circstp
TYPE intern
DESIGN_CLASS RS
TPG_METHOD R
SELF_TEST y
SEQ_DEPTH $SDEPTH
OV_G ($DFF-$OUTPINS)*4-$INPINS*1.5
ACCESS_IN_D y
ACCESS_IN_CO y
ACCESS_IN_CL y
ACCESS_OUT y
ACCESS_BUS y
PERF 1
OR 1
OV_INPIN 3
OV_OUTPIN 1
OV_BIPIN 0
PIN_COMP 1
D_STYLE F
FUN 0
ACH_FC 0.99
NUM_TP (2^$INPINS)*8
CELLS $DFF*4+$INPINS*2.5
```

HFC: Self-testable PLA-design with High Fault Coverage

DESCRIPTION:
Self-testable PLA design based on crosspoint counting.

TEST VECTORS:
product_lines*(2*inputs+outputs+3)+outputs+2

EXTRA GATES:
Extra Transistors:
(12*inputs+6*(product_lines+outputs)+19*(ln(product_lines)/ln(2)))

EXTRA PINS:
2 inputs
1 output

PERFORMANCE IMPACT:
1 gate delay

REFERENCE:
[Saluja85]

NAME hfc
TYPE intern
DESIGN_CLASS PL
TPG_METHOD N
SELF_TEST y
SEQ_DEPTH $SDEPTH
OV_G (12*$INPINS+6*($DFF+$OUTPINS)+19*(ln$DFF/ln2))/4
ACCESS_IN_D n
ACCESS_IN_CO n
ACCESS_IN_CL n
ACCESS_OUT n
ACCESS_BUS n
PERF 1.33
OR 0
OV_INPIN 2
OV_OUTPIN 1
OV_BIPIN 0
PIN_COMP 0
D_STYLE A
D_STYLE F
D_STYLE L

FUN 0
ACH_FC 1
NUM_TP $DFF*(2*$INPINS+$OUTPINS+3)+$OUTPINS+2
CELLS 3+ln($DFF/ln2)

DAEHN: Using bilbo for partitioning

DESCRIPTION:
This self-test method uses BILBO registers placed after input decoders, the AND array, and the OR array.

TEST VECTORS:
2*inputs+product_lines+outputs+3

EXTRA GATES:
(2*inputs+product_lines+maximum (inputs, outputs))*11.5

EXTRA PINS:
2 inputs
1 output

PERFORMANCE IMPACT:
None.

REFERENCE:
[Daehn81]

NAME daehn
TYPE intern
DESIGN_CLASS PL
TPG_METHOD E
SELF_TEST y
SEQ_DEPTH 0
OV_G (2*$INPINS + $DFF + ($INPINS>$OUTPINS) * $INPINS + ($INPINS<$OUTPINS) * $OUTPINS + ($INPINS=$OUTPINS) * $OUTPINS)*11.5
ACCESS_IN_D y
ACCESS_IN_CO y
ACCESS_IN_CL n
ACCESS_OUT y
ACCESS_BUS n
PERF 1
OR 0
OV_INPIN 2
OV_OUTPIN 1

OV_BIPIN 0
PIN_COMP 256
D_STYLE F
D_STYLE L
D_STYLE A
FUN 0
ACH_FC 0.99
NUM_TP 2*$INPINS+$DFF+$OUTPINS+3
CELLS 3

EXHAUSTIVE TEST OF PLA

DESCRIPTION:
All possible input combinations are applied to the PLA as test pattern.

TEST VECTORS:
The number of test vectors is 2^n, where n is the number of inputs.

EXTRA GATES:
No extra gates.

EXTRA PINS:
No extra pins.

PERFORMANCE IMPACT:
None.

REFERENCE:
None.

NAME pla_exh
TYPE intern
DESIGN_CLASS PL
TPG_METHOD E
SELF_TEST n
SEQ_DEPTH $SDEPTH
OV_G 0
ACCESS_IN_D n
ACCESS_IN_CO n
ACCESS_IN_CL n
ACCESS_OUT n
ACCESS_BUS n
PERF 1
OR 1

OV_INPIN 0
OV_OUTPIN 0
OV_BIPIN 0
PIN_COMP 0
D_STYLE A
FUN 0
ACH_FC 1
NUM_TP 2^$INPINS
CELLS 0

TREUER: BIST that uses cumulative parity comparisons

DESCRIPTION:
This method checks the product line using cumulative parity comparison. The parity signals are accumulated in a parity counter and the value is compared with expected values at specific times.

TEST VECTORS:
The number of test vectors is:
(1.25*outputs+3)*(2*inputs*product_lines+2*product_lines+1)

EXTRA GATES:
Extra transistors:
(11*inputs+6*product_lines+3.5*outputs+30)

EXTRA PINS:
2 inputs
2 outputs

PERFORMANCE IMPACT:
1 gate delay

REFERENCE:
[Treuer85]

NAME treuer
TYPE intern
DESIGN_CLASS PL
TPG_METHOD N
SELF_TEST y
SEQ_DEPTH 0
OV_G (11*$INPINS+6*$DFF+3.5*$OUTPINS+30)/4
ACCESS_IN_D n
ACCESS_IN_CO n

ACCESS_IN_CL n
ACCESS_OUT n
ACCESS_BUS n
PERF 1.33
OR 0
OV_INPIN 2
OV_OUTPIN 2
OV_BIPIN 0
PIN_COMP 32768
D_STYLE A
D_STYLE F
D_STYLE L
FUN 0
ACH_FC 1
NUM_TP (1.25*$OUTPINS+3)*(2*$INPINS*$DFF+2*$DFF+1)
CELLS 2+$OUTPINS+10

FUJIWARA: PLA WITH UNIVERSAL TESTS

DESCRIPTION:
The method uses parity checking by using a universal (function independent) test set.

TEST VECTORS:
Number of test pattern:
(3+2*(log (product_lines))/(log (2)))*2*inputs+3*product_lines

EXTRA GATES:
Extra transistors:
(4*inputs+11*product_lines+5*outputs+6)

EXTRA PINS:
3 inputs
2 outputs

PERFORMANCE IMPACT:
1 gate delay

REFERENCE:
[Fujiwara81]

NAME fujiwara
TYPE intern
DESIGN_CLASS PL
TPG_METHOD N

SELF_TEST y
SEQ_DEPTH 0
OV_G (4*$INPINS+11*$DFF+5*$OUTPINS+6)/4
ACCESS_IN_D n
ACCESS_IN_CO n
ACCESS_IN_CL n
ACCESS_OUT n
ACCESS_BUS n
PERF 1.33
OR 0
OV_INPIN 3
OV_OUTPIN 2
OV_BIPIN 0
PIN_COMP 32768
D_STYLE F
D_STYLE L
D_STYLE A
FUN 0
ACH_FC 1
NUM_TP (3+2*(ln$DFF)/(ln2))*2*$INPINS+3*$DFF
CELLS 3

Application of march pattern

DESCRIPTION:
Special test pattern type, writing the complement test pattern and reading in inverse order. Can be implemented as a self-test method (march_self)

TEST VECTORS:
(2*ram_size/data_inpins)+(8*ram_size)
Comment: ram_size in number of bits.

EXTRA GATES: no extra logicfor march, extra gates for self test

EXTRA PINS: no extra pins for march, 8 extra pins for self test.

PERFORMANCE IMPACT:
None.

NAME march	NAME march_self
TYPE intern	TYPE intern
DESIGN_CLASS RA	DESIGN_CLASS RA
TPG_METHOD N	TPG_METHOD N

SELF_TEST n
SEQ_DEPTH $SDEPTH
OV_G 0
ACCESS_IN_D n
ACCESS_IN_CO n
ACCESS_IN_CL n
ACCESS_OUT n
ACCESS_BUS n
PERF 1
OR 1
OV_INPIN 0
OV_OUTPIN 0
OV_BIPIN 0
PIN_COMP 1
D_STYLE F
D_STYLE L
FUN 0
ACH_FC 1
NUM_TP
(2*$RAM_SIZE/$DATAPINS)+(8*$RAM_SIZE)
CELLS 0

SELF_TEST y
SEQ_DEPTH $SDEPTH
OV_G
2.5*$DATAPINS+3*((ln($RAM_SIZE/$DATAPINS))/(ln2))+8.5*$DATAPINS+2.5*($DATAPINS-1)
ACCESS_IN_D y
ACCESS_IN_CO y
ACCESS_IN_CL n
ACCESS_OUT y
ACCESS_BUS y
PERF 1
OR 1
OV_INPIN 7
OV_OUTPIN 1
OV_BIPIN 0
PIN_COMP 1
D_STYLE F
D_STYLE L
FUN 0
ACH_FC 1
NUM_TP
(2*$RAM_SIZE/$DATAPINS)+(8*$RAM_SIZE)
CELLS 0

MARINESCU: TEST METHOD FOR RAMS

TEST VECTORS
ram_size/(data_inputs)+16*ram_size
Comment: ram_size in bits

EXTRA GATES:
2.5*data_inpins+3*((ln(ram_size/data_inpins))/(ln(2)))+6*data_inpins + 2.5*(data_inpins-1)+2*(3*data_inpins+1)

EXTRA PINS:
5 inputs
2 outputs

NAME marinescu
TYPE intern
DESIGN_CLASS RA
TPG_METHOD N
SELF_TEST n
SEQ_DEPTH $SDEPTH

OV_G
2.5*$DATAPINS+3*((ln($RAM_SIZE/$DATAPINS))/(ln2))+6*$DATAPINS+2.5*(
$DATAPINS-1)+2*(3*$DATAPINS+1)
ACCESS_IN_D y
ACCESS_IN_CO y
ACCESS_IN_CL n
ACCESS_OUT y
ACCESS_BUS y
PERF 1.5
OR 1
OV_INPIN 5
OV_OUTPIN 2
OV_BIPIN 0
PIN_COMP 1
D_STYLE F
D_STYLE L
FUN 0
ACH_FC 1
NUM_TP $RAM_SIZE/($DATAPINS)+16*$RAM_SIZE
CELLS 0

ILLMAN: Self-Testing RAMS

DESCRIPTION:
Random pattern application by using standard linear feedback shift registers.

TEST VECTORS:
(ln(1-required_fault_coverage)/ln(0.75))*3*ram_size/data_inputs
Comment: ram_size in number of bits.

EXTRA GATES:
2*(3*(data_inputs+2))+3*((ln(ram_size/data_inputs))/(ln(2))+2)
+2.5*(data_inputs-1)+9*((ln((ln(1-required_fault_coverage)/ln(0.75))
*((ram_size/data_inputs) + (ln(ram_size/data_inputs)))/(ln2)))/(ln(2))+2)
+4+2.5

EXTRA PINS:
4 inputs
2 outputs

REFERENCE:
[Illman86]

NAME illman

TYPE intern
DESIGN_CLASS RA
TPG_METHOD R
SELF_TEST y
SEQ_DEPTH $SDEPTH
OV_G
2*(3*($DATAPINS+2))+3*((ln($RAM_SIZE/$DATAPINS))/(ln2)+2)+2.5*($DATAPINS-1)+9*((ln((ln(1-$FC)/ln(0.75))*(($RAM_SIZE/$DATAPINS)+(ln($RAM_SIZE/$DATAPINS)))/(ln2)))/(ln2)+2)+4+2.5
ACCESS_IN_D y
ACCESS_IN_CO y
ACCESS_IN_CL n
ACCESS_OUT y
ACCESS_BUS n
PERF 1+0.1*(ACC_IN+ACC_OUT)/($INPINS+$OUTPINS)
OR 1-(ACC_IN+ACC_OUT)*18/$GATE
OV_INPIN 4
OV_OUTPIN 2
OV_BIPIN 0
PIN_COMP 1
D_STYLE F
D_STYLE L
FUN 1
ACH_FC $FC
NUM_TP (ln(1-$FC)/ln(0.75))*3*$RAM_SIZE/$DATAPINS
CELLS 0

ROM_NODFT: EXHAUSTIVE TEST OF ROM

DESCRIPTION:
All storage cells are read and compared with expected fault-free values.

TEST VECTORS:
The number of test vectors is n, where n is the number of storage cells.

EXTRA GATES:
No extra gates.

EXTRA PINS:
No extra pins.

PERFORMANCE IMPACT:
None.

NAME rom_nodft
TYPE intern
DESIGN_CLASS RO
TPG_METHOD E
SELF_TEST n
SEQ_DEPTH $SDEPTH
OV_G 0
ACCESS_IN_D n
ACCESS_IN_CO n
ACCESS_IN_CL n
ACCESS_OUT n
ACCESS_BUS n
PERF 1
OR 1
OV_INPIN 0
OV_OUTPIN 0
OV_BIPIN 0
PIN_COMP 0
D_STYLE F
D_STYLE L
FUN 0
ACH_FC 1
NUM_TP $DFF
CELLS 0

EXT_SCAN_SET: External Scan-Set-Design

DESCRIPTION:
A scan-set element is put at every input and every output of a block. All latches are connected to a scan register. Full accessibility of the inputs and outputs of the block is achieved.

EXTRA GATES:
(inputs+outputs)*size_of_a_scan_set_element

EXTRA PINS:
2 inputs
1 output

PERFORMANCE IMPACT:
1 clock-cycle per pin.

REFERENCE:
[Williams83]

NAME ext_scan_set
TYPE extern
DESIGN_CLASS RS
DESIGN_CLASS RC
DESIGN_CLASS PL
DESIGN_CLASS RA
DESIGN_CLASS RO
DESIGN_CLASS MU
TPG_METHOD S
SELF_TEST n
SEQ_DEPTH $SDEPTH
OV_G (ACC_IN+ACC_OUT)*18+$ACC_BUS*12
ACCESS_IN_D y
ACCESS_IN_CO y
ACCESS_IN_CL n
ACCESS_OUT y
ACCESS_BUS y
PERF 1+0.1*(ACC_IN+ACC_OUT)/($INPINS+$OUTPINS)
OR 1-(ACC_IN+ACC_OUT)*18/$GATE
OV_INPIN 3
OV_OUTPIN 1
OV_BIPIN 0
PIN_COMP 1
D_STYLE A
D_STYLE F
D_STYLE L
FUN 0
NUM_TP C
ACH_FC C
CELLS 0

EXT_D_LATCH: External double latch

DESCRIPTION:
A double-latch is put at every input and every output of a block. All latches are connected to a scan register. Full accessibility of the inputs and outputs of the block is achieved.

EXTRA GATES:
(inputs+outputs)*size_of_a_double_latch

EXTRA PINS:
3 inputs
1 output

PERFORMANCE IMPACT:
2 gate delays per pin.

REFERENCE:
[Williams83]

NAME ext_d_latch
TYPE extern
DESIGN_CLASS RS
DESIGN_CLASS RC
DESIGN_CLASS PL
DESIGN_CLASS RA
DESIGN_CLASS RO
DESIGN_CLASS MU
TPG_METHOD S
SELF_TEST n
SEQ_DEPTH $SDEPTH
OV_G (ACC_OUT+ACC_IN)*7+$ACC_BUS*10
ACCESS_IN_D y
ACCESS_IN_CO y
ACCESS_IN_CL n
ACCESS_OUT y
ACCESS_BUS y
PERF 1.1
OR 0.95
OV_INPIN 3
OV_OUTPIN 1
OV_BIPIN 0
PIN_COMP 2
D_STYLE A
D_STYLE F
D_STYLE L
FUN 0
NUM_TP C
ACH_FC C
CELLS 0

References

[Abadir85] Abadir, M.S., Breuer, M.A., 'A Knowledge Based System for Designing Testable VLSI Chips', IEEE Design and Test of Computers, Vol. 2, No 4, August 1985, pp. 56-68

[Abadir89] Abadir, M.S., 'TIGER: Testability Insertion Guidance Expert System', Proc. IEEE International Conference on Computer Aided Design, 1989, pp. 562-565.

[Abramovici90] Abramovici M., Breuer M.A., Friedman, *Digital Systems Testing and Testable Design*, Computer Science Press, 1990.

[Agrawal87] Agrawal, V.D., Cheng, K.T., Johnson, D.D., Lin, T., 'A Complete Solution to the Partial Scan Problem', Proc. IEEE International Test Conference, 1987, pp. 44-51.

[Ambler86] Ambler, A.P., Paraskeva, M., Burrows, D.F., Knight, W.L., Dear, I.D., 'Economically Viable Automatic Insertion of Self Test Features for Custom VLSI', IEEE International Test Conference, 1986, pp. 232-243.

[Ando79] Ando, H., 'Testing VLSI with Random Access Scan', IEEE International Test Conference, 1979, pp. 37-41.

[Ardeman87] Ardeman, A., 'ASICs: Costing and Planning', New Electronics, 20 January 1987.

[Beenker86] Beenker, F.P.M., van Eerdewijk, K.J.E., Gerritse, R.B.W., Peacock, F.N., van der Star, M., 'Macro Testing: Unifying IC and Board Test', IEEE Design and Test of Computers, December 1986.

[Beenker90] Beenker, F., Dekker, R., Stans, R., van der Star, M., 'Implementing Macro Test in Silicon Compiler Design', IEEE Design and Test of Computers, April 1990, pp. 41-51.

[Bennetts80] Bennetts, R.G., Maunder, C., Robinson, G., 'CAMELOT: A computer-aided measure for logic testability', Proceedings IEEE International Conference on Computer Design, 1980, pp. 1162-1165.

[Bennetts89] Bennetts, R.G., 'Definition of basic terms used in testing, and a list of abbreviations used in WP5', ESPRIT - EVEREST report no. BEN0026 TN 01 IN W5.

[Bennetts92] Bennetts, R.G Keynote address, IEEE International Test Conference, 1992.

[Bhavsar86] Bhavsar, D.K., 'A New Economical Implementation for Scannable Flip-Flops in MOS', IEEE Design and Test of Computers, Vol. 3, No 3, June 1986, pp. 52-56.

[Blanchard78] Blanchard, B.S., *Design and Manage to Life Cycle Cost,* M/A Press, 1978

[Blohm88] Blohm, H., Lueder, K., *Investition,* Verlag Vahlen, 6th edition, 1988.

[Blumberg84] Blumberg, D., 'Remote Diagnostics Seeks to Aid Field Services', Data Communication, August 1984.

[Blumberg90] Blumberg, D., 'Strategies for Improving Productivity in Field Service Organisations', AFSM International, November 1990, pp. 14-33.

[Breuer85] Breuer, M.A, Zhu, X., 'A Knowledge Based System for Selecting a Test Methodology for a PLA', Proc. 22nd IEEE Design Automation Conference, pp. 259-265.

[Breuer88] Breuer, M.A., Gupta, R., 'Knowledge based systems for test and diagnosis', in *Knowledge Based Systems for Test and Diagnosis*, Saucier, Ambler, Breuer (eds), North Holland, 1989, pp. 3-27.

[Bryant89] Bryant, W., 'Logistics Impacts and Influences of ATE Design to Meet the TLM Concept', Proceedings of Autotestcon89, IEEE, 1989, p. 32.

[CAIS90] CAIS, Technology Strategies, May 1990, pp. 1-15.

[Cheng89] Cheng, K. T., Agrawal. V.D., 'A Partial Scan Method for Sequential Circuits with Feedback', IEEE Transactions on Computers, vol. 39, no. 4, April 1989, pp. 544-548.

[Comerford82] Comerford, R., 'Automation Promises to Lighten the Field Service Load', Electronics, April 1982.

[Croft85] Croft, R.M., Baker, K., 'Global Testability Tool for CATE', Proc. IEE Colloquium on Design for Testability, Nov. 85.

[Daehn81] Daehn, W., Mucha, J., 'A Hardware Approach to Logic Testing of Large Programmable Logic Arrays', IEEE Trans. on Computers, vol. C-30, November 1981, pp. 829-833.

[Davis82] Davis, B., *The Economics of Automatic Testing*, McGraw-Hill, 1982

[Davis92] Davis, B., 'The economics of design and test', in Proc. Economics of Design and Test for Electronic Circuits and Systems, Ellis Horwood, 1992

[Davis93] Davis, B., 'Economic Modelling of Board Test Strategies', 2nd International Workshop on the Economics of Design, Manufacturing and Test, Austin, May 1993.

[Davis94] Davis, B., *The Economics of Automatic Testing*, McGraw-Hill, 1994.

[Dear87] Dear, I.D., Ambler, A.P., 'Predicting Cost and Quality Improvements as a Result of Test Strategy Planning', IFIP Workshop on Fast Prototyping, March 1987.

[Dear89] Dear, I.D., Dislis, C., Miles, J.R, Lau, S.C, Ambler, A.P., 'Hierarchical Testability Measurement and Design for Test Selection by Cost Prediction', European Test Conference 1989, Paris, pp. 50-57.

[Dervisoglu88] Dervisoglu, B.I., 'Using Scan Technology for Debug and Diagnostics in a Workstation Environment', IEEE International Test Conference, Washington, USA, September 1988.

[Deway86] Deway, A., Gadient, T., 'VHDL Motivation', IEEE Design and Test of Computers, April 1986, pp. 12-16.

[Di Giacomo89] Di Giacomo, J., *VLSI Handbook.*, McGraw Hill publishing Company, 1989.

[Dinkelbach69] Dinkelbach, W., *Sensitivitätsanalysen und parametrische Programmierung*, Springer Verlag, 1969

[Dislis89] C Dislis, I D Dear, J R Miles, S C Lau & A P Ambler, 'Cost Analysis of Test Method Environments', IEEE International Test Conference, 1989, pp 875-883.

[Dislis91] Dislis, C., Dick, J., Ambler, A.P., 'An Economics Based Test Strategy Planner for VLSI Design', Proceedings of the European Test Conference, 1991.

[Dislis93] Dislis, C., Dick, J.H., Azu, I.N., Ambler, A.P., 'Economics Modelling for the Determination of Test Strategies for Complex VLSI Boards', Proc. IEEE International Test Conference, 1993.

[Eichelberger77] Eichelberger, E.B., Williams, T.W., 'A Logic Design Structure for LSI Testability', Proc. 14th Design Automation Conference, 1977, pp. 462-468.

[Fey87] Fey, C.F., Paraskevopoulos, D.E., 'A Techno-Economic Assessment of Application-Specific Integrated Circuits: Current Status and Future Trends', Proc. of the IEEE, vol. 75, no. 6, June 1987, pp. 829-841.

[Fey88a] Fey, C.F., Paraskevopoulos, D.E., 'Economic Aspects of Technology Selection: Level of Integration, Design Productivity, and Development Schedules', chapter in *VLSI Handbook,* Di Giacomo, McGraw-Hill, 1988.

[Fey88b] Fey, C.F., Paraskevopoulos, D.E., 'Economic Aspects of Technology Selection: Costs and Risks', chapter in *VLSI Handbook,* Di Giacomo, McGraw-Hill, 1988.

[Fink84] Fink, G., 'On-Board Diagnostics Lower Service Costs for OEMs and System Integration', Mini-Micro systems, May 1984.

[Fujiwara81] Fujiwara, H., Kinishita, 'A design of Programmable Logic Arraya with universal tests', IEEE Transactions on Computers, Vol. c-30, November 1981, pp. 823-828.

[Fujiwara83] Fujiwara, H., Shimono, T., 'On the Acceleration of Test Generation Algorithms', IEEE Transactions on Computers, vol. C-32, 1983.

[Fuller88] Fuller, M., 'How Design for Service Speeds Production Development Time and Reduces Service Costs', ATE and Instrumentation Conference East, 1988, pp. 549-554.

[Fung86] Fung, H.S., Hirschhorn, S., 'An Automatic DFT system for the Silk Silicon Compiler', IEEE Design and Test of Computers, February 1986, pp. 45-57.

[Gebotys88] Gebotys, C.H., Elmasry, M.I., 'Integration of Algorithmic VLSI Synthesis With Testability Incorporation', Proc. IEEE Custom Integrated Circuits Conference, 1988, pp. 2.4.1-2.4.4.

[Goel80] Goel, P., 'Test Generation Costs Analysis and projections', Proc. IEEE 17th Design Automation Conference, 1980, pp 77-84.

[Goel81] Goel, P., 'An Implicit Enumeration Algorithm to Generate Tests for Combinational Logic Circuits', IEEE Transactions on Computers, C-30, 1981

[Goldstein80] Goldstein, L.H., Thigpen, E.L., 'SCOAP: SANDIA Controllability/Observability Analysis Program', 17th Design Automation Conference, June 1980, pp. 180-186.

[Gomez91] Gomez, D., IEEE Design and Test Roundtable, June 1991.

[Gundlach90] Gundlach, H.H.S., Müller-Glaser, K.D., 'On Automatic Test Point Insertion in Sequential Circuits', Proc. IEEE International Test Conference, 1990, pp. 1072-1079.

[Hammersley65] Hammersley, J.M., Handscomb, D.C., *Monte Carlo Methods*, Methuen, London, 1965.

[Hills85] Hills, T.G., Davis, M.J., Rogers, W.E., 'Cost Visibility Through Life Cycle Cost Boards', Logistics Spectrum, vol. 19, Los Angeles 1985, pp. 23-25.

[Huber91] Huber, J.P., Rosneck M., *Successful ASIC Design the First Time Through*, Van Nostrand Reinhold, 1991.

[Hughes89] Hughes, W., 'Two Levels vs Three Levels of Maintenance: The Cost', Autotestcon89, IEEE 1989, pp. 19-25.

[Huston83] Huston, R.E., 'An Analysis of ATE Testing Costs', IEEE International Test Conference 1983, pp. 396-411.

[Illman86] Illman, R.J., 'Design of a Self-Testing RAM', Proc. Silicon Design Conference 1986, pp. 439-446.

[Intelligen89] Racal-Readac, *Intelligen Manual,* 1989.

[Jones86] Jones, N.A., Baker, K., 'An Intelligent Knowledge Based System Tool for High Level BIST Design', IEEE International Test Conference, September 1986, pp. 743-749.

[Kirk79] Kirk, S.J., 'Life Cycle Costing: Problem Solver for Engineers', Specifying Engineer, Washington D.C. 1979, pp. 123-129.

[Kirkpatrick83] Kirkpatrick, S., Gelatt, C.D., Vecchi, M.P., 'Optimisation by Simulated Annealing', Science, Vol 220 1983, pp. 671-680.

[Kitchenham90] Kitchenham, B., de Neumann, B., 'Cost Modelling and Estimation', chapter 11, *Software Reliability Handbook*, Paul Rook (ed.), Elsevier Sciaence Publishers, 1990, pp. 333-376.

[Koenemann79] Koenemann, B., Mucha, J., Zwiehoff, G., 'Built-in Logic Block Observation Techniques', IEEE International Test Conference, 1979, pp. 37-41.

[Krasniewski88] Krasniewski, A., Pilarski, 'Circular Self Test Path - A Low Cost BIST Technique', Proc. IEEE 24th Design Automation Conference, 1988, pp. 3-8.

[Levitt89] Levitt M.E, Abraham J.A., 'The Economics of Scan Design', IEEE International Test Conference, August 1989.

[Luenberger73] Luenberger, D.G., *Introduction to Linear and Nonlinear Programming, Addison Wesley*, 1973

[Madauss84] Madauss, B.J., *Projektmanagement*, 2nd edition, Stuttgart 1984.

[McCluskey85] McCluskey, M.J., 'Built-in Self Test Techniques', IEEE Design and Test of computers, April 1985, pp 21-27

[Miles91] Miles, J., De Bondt, R., Daemen, L., 'A Test Economics Model and Cost Benefit Analysis of Boundary Scan', Proc. European Test Conference, 1991.

[Murray88] Murray, W., 'Is Two-Level Repair Feasible', ATE and Instrumentation Conference East, 1988, pp. 571-681.

[Myers83] Myers, M.A., 'An Analysis of the Cost and Quality Impact of LSI/VLSI Technology on PCB Test Strategies', Proc. IEEE International Test Conference, 1983, pp. 382-395.

[Needham91] Needham W., *Designer's Guide to Testable ASIC Devices*, Van Nostrand Reinhold, 1991.

[Ohletz87] Ohletz, M.J., Williams, T.W., Mucha, J.P., 'Overhead in Scan and Self-Testing Designs', Proc. IEEE International Test Conference, 1987, pp. 460-470.

[Perone86] Perone, J., 'Reducing Test Costs through Strategic Changes in Maintenance and Service', IEEE International Test Conference, 1986, pp. 205-212.

[Pynn86] Pynn, C., *Strategies for Electronics Test*, McGraw-Hill, 1986.

[Racal89] Racal-Redac Seminar, 1989.

[Reinertsen83] Reinertsen, D.G., 'Whodunit? The Search for the New Product Killers', Electronic Business, Vol. 11, July 1983, pp. 106 - 109.

[Rubinstein86] Rubinstein, R.Y., *Monte Carlo Optimization, Simulation and Sensitivity of Queueing Networks*, John Wiley & Sons, 1986

[Saluja85] Saluja, Upadhaya, 'A Built-in Self-Test PLA Design with Extremely High Fault Coverage', Technical report EE 8517, Department of Electrical and Computing Engineering, University of Newcastle, NSW, Australia, March 1985.

[Samad86] Samad, M.A., Fortes, J.A.B., 'Explanation Capabilities in DEFT - A Design-for-Testability Expert System', Proc. IEEE International Test Conference, 1986, pp. 954-963.

[Sedmak92] Sedmak, R., 'Economics and Other Management Issues', Design for Testability Training Course, held at Texas Instruments, Munich, 1992.

[Smith80] Smith, J.E., 'Measure of the Effectiveness of Fault Signature Analysis', IEEE Transactions on Computers, Vol C-29, No.6, June 1980.

[Smith91] Smith, P.G., Reinertsen, D.G., 'Developing products in half the time', Van Nostrand Reinhold, 1991.

[Szygenda92] Szygenda, S.A., 'Profit, Liability, and Education: Influencing Factors on the Economics of Non-Testing', in *Proc. Economics of Design and Test for Electronic Circuits and Systems*, Ellis Horwood, 1992.

[Treuer85] Treuer, R., Fujiwara, H., Agrawal, V.K., 'Implementing a Built-in Self-Test PLA Design', IEEE Design and Test of Computers, Vol. 2, April 1985.

[Trischler83] Trischler, E., 'Testability Analysis and Incomplete Scan Path', Proc. IEEE International Conference on Computer Aided Design, 1983, pp. 38-39.

[Turino90] Turino, Jon, *Design to Test - A definitive guide for electronic design, manufacture, and service*, Van Nostrand Reinhold, New York, 1990.

[Varma84] Varma, P., Ambler, A.P., Baker, K., 'An Analysis of the Economics of Self-Test', Proc. IEEE International Test Conference, 1984.

[Warwick84] Warwick, W., keynote address at IEEE International Test Conference 1984, published in IEEE Design and Test of Computers, November 1984.

[Wilkins86] Wilkins B.R., *Testing Digital Circuits: An Introduction*, Van Nostrand Reinhold, 1986.

[Williams83] Williams, T.W., Parker, K., 'Design for Testability - A survey', Proceedings of IEEE, Vol 71, No. 1, January 1983, pp. 98-112.

[Wong88] Wong, D.F., Leong, H.W., Liu, C.L., 'Simulated Annealing for VLSI Design', Kluwer Academic Publishers, 1988.

[Wübbenhorst84] Wübbenhorst, K., *Konzept der Lebenszykluskosten*, Verlag für Fachliteratur, 1984.

[Zhu88] Zhu, X., Breuer, M.A., 'Analysis of Testable PLA Designs', IEEE Design and Test of Computers, Vol. 5, No. 4, pp. 14-28.

Index

—O—

—P—

—Q—

—R—

—S—

—T—